Cultivating
Carnivorous Plants

Allan A. Swenson

DOUBLEDAY & COMPANY, INC.
GARDEN CITY, NEW YORK
1977

Photos by the author
and Peter Jon Swenson
unless otherwise noted.

Illustrations by Donna R. Sabaka

Library of Congress Cataloging in Publication Data

Swenson, Allan A
    Cultivating carnivorous plants.

    Bibliography: p. 152
    Includes index.
    1. Insectivorous plants.   I. Title.
SB432.S9   635.9′33′121
ISBN 0-385-11148-7
Library of Congress Catalog Card Number 75–36614

William S. Northrop (1901–1971), a self-taught naturalist, horticulturist, botanist, biologist, philosopher, and friend, to whom I am truly indebted.

Acknowledgments

For twenty-plus years I have lived among carnivorous plants. During that time I have traveled extensively to study these plants, photograph them, and interview scientists and amateur growers alike who share my fascination with these botanical wonders.

In greenhouse, on windowsill, in office, studio, even in underground artificial light experimental rooms, I have sought to unravel some of the puzzles surrounding these strange plants and their astonishing habits. There have been many long days and nights perched among these plants.

I am grateful for and willingly acknowledge the help provided by my entire family: my understanding wife, Sheila, and our sons Peter, Drew, Boyd, and Meade, who have helped so many hours in the growing, care, and study of these plants. Peter especially has aided immeasurably in the research phases.

Many others too contributed their thoughts and ideas as well as knowledge to this book. In particular I must thank Mrs. William Northrop, George and June Tregembo, and the legions of teachers, students, botanists, and other writers who have given their time, assistance, and experience in the creation of this work. I am sincerely grateful to all of them.

Contents

Introduction

Scattered through the pages of history, in fable and folklore, strange tales of wild and wicked plants abound. As the stories go, these terrifying plants are capable of snatching animal victims and greedily devouring them.

Do such carnivorous wonders of the plant world really exist? Are the fantastic reports that periodically have seeped out from the jungles of the Amazon, the hidden valleys of Indonesia, and the hot, dark heart of Africa just legends? Do gigantic plants that thrive by eating animals and birds at will still thrive in some remote, unexplored parts of our planet?

I've always been curious about such things. After all, not long ago a tribe of Stone Age natives—the Tasaday—that should not still exist was discovered in a remote jungle of the Philippines. Perhaps there are some plants, left-over relics of the long-gone dinosaur days, still lurking somewhere.

After more than twenty years of traveling the world, corresponding, researching, studying, and writing about carnivorous plants, I must give a qualified "yes" when asked if such strange plants really do exist today. There are in fact many truly exotic and bizarre meat-eating plants lurking in distant corners of the globe. Some, not so very far away.

The largest known species can actually catch and devour birds, rodents, and similar unsuspecting prey. Not only can they, do. Among the nearly five hundred types of carnivorous plants already known, some have an exceptional and surprising ability to lure, catch, and digest their insect, animal, and fish dinners.

Fortunately for us all, despite my best efforts this past quarter century, I have never found those man-eating "monster" plants so popular with science fiction writers.

Perhaps in other times, imaginative tellers of tall tales found eager ears for their early versions of science fiction stories. The plants they described so vividly actually do exist—though not nearly as exaggerated as the stories made them out to be—quite capable of snaring unwary insects with sticky tentacles, swiftly snapping shut to trap a small fly or moth or frog, or drowning their victims in their soupy broth.

Fact is, in all my travels and years of study of this wide range of botanical wonders, I haven't met a carnivorous plant I didn't like. Not yet, anyway.

Even so talented a scientist as the famed naturalist Charles Darwin was fascinated by these botanical wonders. Among them, one stands out. Darwin himself described the Venus flytrap, that can snap shut to catch its insect meal in a split second, as "the most wonderful plant in the world."

His experiments with the flytrap and others with similar carnivorous habits remain the foundation for all future studies of these plants.

There are hundreds of plants that have this unique and amazing ability to lure, catch, and eat insects and small animals. Some can even manage a small bird or two as part of their diet.

Today, many of these fascinating plants can be domesticated. You too can marvel at their feats when you grow them as house plants, in terrariums or as conversation-provoking plant pets.

Actually, a number of the larger carnivorous plants can be cultivated quite easily. Your friends will be amazed at their versatility, not to mention their agility. After all, how many times do they see an entire collection of plants merrily devouring gnats, flies, moths, mosquitoes, and other forms of animal life.

Whether you just want to know a bit about these legendary meat-eating plants or plan some serious studies of their curious abilities, you can enjoy hours, even years of pleasure with the most wonderful plants in the world.

To help you with your studies or just plain fun gardening with these captivating plants, I have gathered as much scientific and practical information as possible. From my years of personal experience, research, and interviews with others involved with carnivorous plants, I hope you will find new horizons in our green growing world.

We have concentrated on those plants which are not only, for most people, the most interesting, but which also are most easily grown. Some types are difficult to obtain. Other, highly unusual species just don't respond well except under special and somewhat difficult-to-maintain conditions.

For those of you with inquisitive minds, we have also compiled suggested study projects, from simple feeding tests to more complex science projects. As you expand your own growing horizons and perfect your horticultural skills, I hope you will add to the needed store of information on this fascinating field of carnivorous plants.

Good luck and good growing.

ALLAN A. SWENSON

Windrows Farm
Kennebunk, Maine

Cultivating
Carnivorous Plants

The Culture and Care of Carnivorous Plants

With few exceptions, carnivorous plants are rather easily grown. If you pay attention to the proper growing medium, the light needs and humidity requirements you can be successful with these wonders of the plant world.

Unlike conventional house plants, carnivorous plants actually require less care. For one thing, they don't want or need fertilizer. That solves a common problem most people have with house plants. For another, they have no need of insecticide sprays. That's good news to organic gardeners as well as anyone who has fought the battle of the bugs around the house plants and in their gardens.

Obviously these carnivorous plants can't abide insecticides. They need those insects. They are the plants' nourishment.

You see, you're far ahead already, since we eliminate some of the problem areas of growing plants—fertilizer and pest control materials. The other ingredients required by regular plants are much more easily provided.

As you read through the various chapters about individual plants and their families of relatives, you will pick up some basic culture tips. This chapter is designed to provide a basic understanding of the plant needs in general and the best ways to provide the special requirements of carnivores to help them thrive and perform perfectly for you.

Containers

First step with carnivorous plants is the container. True, many can be grown successfully as potted plants. But here's a warning. Because carnivorous plants require lots of water to satisfy their needs for secreting insect-attractant aromas, digestive fluids, and providing the pressures for their snapping or closing, wrapping or folding actions, adequate moisture is essential.

Luckily, terrariums have made a comeback. Carnivorous plants really enjoy life in terrariums. But as these plants need mobile food, we can't grow them in totally closed terrariums. How would the insects enter? So, we're limited a bit, but not much.

Several years back you may have been limited in your container choice to fish tanks and bowls, gallon jars, and other less-than-attractive planting containers. Today, with the boom in terrarium culture, you have a far-ranging selection of containers. They extend from beautiful

A Crystal Lite Greenhouse unit with twin-tube fluorescent light fixture makes an ideal high-humidity terrarium for a collection of carnivorous plants.

brandy snifters that will hold several plants to complete glass-topped terrarium tables. You can choose small Tiny Terras and Gro Dome planters for individual plants or grow complete collections in Crystal Lite indoor greenhouses.

Which containers you select depends on your own personal preferences. You naturally want to match the type containers to your decorating scheme. Eye appeal is important.

Most important to your carnivorous pets, however, is a container that

When filled with moss, brandy snifters like these 32-ounce glasses make excellent planters for carnivorous plants. That's a Venus flytrap on the right; the other plant is a purple pitcher.

will insure the higher humidity they require. Glass planters usually prove best. They are easily cleaned, provide excellent viewing from all angles, and do hold that important humidity around the plants. New designs in plastic terrariums now provide even wider choices.

Many homes unfortunately are overheated, but as fuel costs go up, temperatures will undoubtedly be lowered, which is to the good for your carnivorous plants. Most of them prefer temperatures from 65° to 75° F. Some tolerate a greater range of temperatures.

The hotter you keep your home, the more moisture is drained for any plants. A hot-air furnace is the most drying type of heating system. That's why semi-enclosed terrariums are desired for growing these unique plants with their special humidity requirements.

For years we have used a variety of growing units. We have seen and photographed others. In all cases, the containers that assume that vital, sustained high humidity will prove the most successful.

You can obtain containers in most garden centers, hardware stores, florist shops, even in supermarkets and glassware stores. From gallon terrarium units to fifty-gallon fish tanks, from apothecary jars of various sizes to expensive plantariums, there is a wide selection of suitable containers.

For the economy-minded, here are some suggestions to keep container costs down so you can spend the savings for additional plants for your collection.

Consider those plastic water and cocktail glasses sold in 5 & 10¢ stores. Upside down over a standard plastic pot they provide a handy and effective high humidity growing chamber.

Large photo cubes from the photo store are reasonable. With the top back in place, or moved slightly to let insects enter, you have a handsome smaller terrarium. Check your local restaurants. They usually purchase olives, cherries, mayonnaise in large-mouth gallon jars. They are excellent for several plants.

Look for tall containers. Many of the most striking pitcher plants grow 15 to 20 inches high. Add a few inches for your planting material and you need a container at least 24 inches tall. True, you can and should leave the top off containers periodically so that potential meals can reach the plants. That means, of course, you'll need to add water more frequently, especially with pitcher plants so they can draw up the liquid they need to digest insects inside their hollow pitchers.

We have seen many interesting and functional plus some quite attractive improvised growing units. One chap simply collected old wooden storm windows, repainted them, and then assembled them into a giant terrarium with hinged top. A variation on that was a teacher who took storm windows and nailed the tops together in A-frame style. The two sides were plastic for easier care.

Plastic sheets, from rigid Plexiglas to the flexible Mylar, or polyethylene bag material can be well used. Just make the size frame you wish, then staple or screw on the material and you have an adequate chamber.

Finding suitable containers is easy. For more extensive collections, you might even prefer a window greenhouse unit which attaches right to the house in place of a storm sash. Or, if you already have a greenhouse you can, with addition of a misting or automatic watering system, provide the desired growing conditions for your carnivorous plants.

Please don't get the idea that these plants are difficult to grow. They aren't. In fact, they are reasonably easy to grow from the bulbs, roots, small plants, and rhizomes which are offered by various suppliers who specialize in these oddities.

We emphasize the need for high humidity to help you realize that attention to this key factor will put you well along the road to success with these particular members of the green world.

This attractive, handy windowsill greenhouse measures only 5¾ inches long, 3¾ inches wide, and 5 inches high. It provides sufficient humidity for carnivorous plants like these flytraps, plus openings for insects to enter.

This decorative glass bowl is an attractive home for a family of blooming butterworts.

Planting Medium

Next basic consideration is your planting medium. Through the years, I have grown tens of thousands of these carnivorous plants, of all sizes and varieties. We've also worked with others who have grown many types: as a hobby, for school projects, and as suppliers to thousands of people who enjoy them around the entire world.

Yes, various combinations of peat moss and sand, vermiculite and peat, perlite mixed with sand and acid soil can be used. But for years of study, and periodic tests of new, potentially better mixes, we still come back to sphagnum moss.

This natural ingredient is a living moss found in every state across the country. It thrives in bogs, along stream banks, in bays of lakes, in swamps and roadside ditches. It is used and sold by florists and many garden centers. Their material is usually dried and supplied to them in large bags or bales.

If you can't find sphagnum moss locally, most carnivorous plant growers supply it with their bulbs and plants anyway. A good source is the Plant Oddities Club, which has specialized in carnivorous plants for years. They provide it with their plants and sell it in small quantities by mail order. You'll find them listed in the sources at the end of this book.

This is the mix we prefer, based on growing tens of thousands of plants over the twenty-five years we have been involved with them.

For flytraps, sundews, butterworts, pitcher plants, and cobra lilies, we use sphagnum moss alone. It has the capacity to hold water, yet is spongy enough so roots can also breathe. It works well for all the plants alone.

For terrarium plantings, we have found that variations also can provide satisfactory results.

If sphagnum moss isn't available, you can grow individual plants, or groups of them for that matter, in the following planting medium or variations of it. By variations, we suggest you try different amounts and combinations of materials with a few plants. If it works for you—if your plants prosper—by all means plant others in it.

Mix one part builder's sand with one part peat. Place about an inch of washed gravel on the bottom of the terrarium or large container. Add several inches of the sand and peat mixture. Then, add another inch or two of sphagnum moss. Those plants that prefer the moss locale will set shallow roots anyway. The deeper-rooted pitchers will use the sand and peat base to steady themselves as they attain their tall maturity.

It is true that in their native habitat these carnivorous plants do grow in various types of soil. Most are acid; that is, they are on the lower numbered part of the pH scale. Since these plants do grow naturally in acid soil, it stands to reason that you should also provide them with a basic natural ingredient, an acid soil.

Well, the acid soil you can obtain locally may be a far cry from the native soil in which the plants originally were grown. Try it if you wish, but it will pay you to mix in good quantities of sphagnum moss with it for insurance. Some soils that look just fine, you see, are composed of too much clay and silt. Roots can't penetrate properly. Air and water can't move well through it. Plant roots can actually drown in soil which is overly wet too long.

So, back we come to sphagnum moss. Try other combinations, from peat and sand to vermiculite and perlite. Experiment. But for insurance, try always to use at least one third sphagnum moss in your planting medium. Your plants will appreciate your thoughtfulness.

When you plant a terrarium or any other house plant, carnivorous or not, it helps to realize that some plants require different conditions than others. Accomplished terrarium enthusiasts realize that they must select compatible plants, those that prefer similar conditions of moisture, heat, light, and humidity if they are to be successful.

Fortunately, carnivorous plants, generally speaking, prefer similar conditions. Some may like a bit more light, others semi-shade. They do, however, prosper together since every plant has a range of tolerance for growing conditions slightly different than what it would ideally prefer.

For individual planting, it also is wise to use a gravel layer before placing plants and sphagnum moss into the container. We've found it helps to place the moss atop any other material, then make a fist-size depression in the center of the container. Wrap a handful or two of moss around the roots and rhizomes of the larger plants. Insert this mossed root ball into depression and top it with a tidy layer of sphagnum. Water well and keep the planting medium moist until the plants have taken hold, set roots, and begun to grow.

For smaller plants, like butterworts, sundews, young pitcher plants, place the plant on top of a handful of sphagnum in your palm. Tuck the plant into the moss, then place it where you wish in the container. With flytrap bulbs, wrap the moss around the bulb and set it into the growing container, with the sprouting leaves just at the surface.

To assemble a total terrarium, add your gravel base first. You may wish to add some charcoal next, about a half inch, which may help absorb odors from excess water. But, if you overwater, which you shouldn't do, the charcoal won't solve the problem of soggy roots.

Keep in mind that roots must breathe, despite the fact that these are plants which grow naturally in or near bogs and swamps. Too much water, like excess anything, can be harmful. Use common sense. Also, watch your plants' responses. If they look ill, pale, or spindly, they're giving you a message. It says "I need better care."

Flytraps, sundews with the exception of intermedia, butterworts, and the smaller parrot pitchers prefer a moist but not constantly soggy area. Pitcher plants, including the illustrious cobra lily, must draw ample moisture from the planting medium to supply that important digestive fluid inside their hollow pitchers.

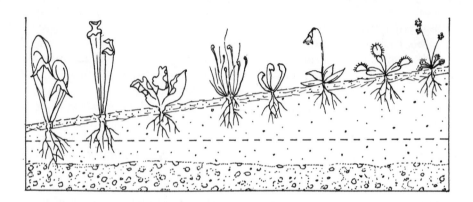

So, slant your terrarium planting to accommodate the needs of these different plants.

Place sundews, flytraps, and butterworts on the higher areas of the terrarium. Plant the pitchers in the lower areas. In this way, those that need extra moisture can get it easily.

You can achieve the same results with a level planting by placing the pitcher plants deeper in the medium. However, we and others have found the slanting method does work better. It is to the advantage of the plants because it caters to their needs.

Once your planting is completed, water well. Use clean stream or brook water, or some obtained from pond or lake. Rain water is excellent. So is well water, unless you are in a hard water area. Excess mineral content in water is detrimental.

If you must use tap water from city supplies that have been treated with chlorine or other chemicals, let it stand for several days. That allows the chlorine to dissipate. Although distilled water has no impurities, it does lack the few useful minor elements which rain, well, or stream water can provide. Of course, if the local streams are polluted, avoid them. No sense adding phosphate detergent pollution or other harmful materials that may damage your valuable plants.

Lighting

Next consideration for your plants is adequate light. Since carnivorous plants range in their requirements from partial or semi-sun to nearly full sun, don't decide to put them in a bright southern exposure window. You can fool yourself and harm them. That rule holds for house plants and terrariums too.

Bright sun for many hours each day will pour through that southern window. Sit in front of it and see how hot it gets. If plants are inside of glass or plastic, even an open container, the heat inside builds up. It becomes too hot for plants. You don't like it ultra hot—neither do your plants.

A good rule of thumb is that plants enjoy temperature ranges which are pleasant for us. East or west windows or a distance away from the southern exposure window is better. Another factor comes into play. In fall and winter and at times in early spring, cold drafts blow through those windows. Plants can stand some temperature extremes but icy drafts should be avoided.

In greenhouses or larger indoor terrariums for which you provide supplemental light, plants will thrive at 60° to 80° F. At higher temperatures they tend to dry.

There are exotic tropical carnivorous plants, like nepenthes, that you might assume prefer the hotter temperatures, bright sun, and other conditions you equate with the equator. Not quite so.

True, those tropical plants can take a bit more heat. But remember, many are native to the lower stories of those tropical forests, as others are to the floor of North American forests and fields. They are protected by taller trees, shrubs, vines. The same is true with flytraps, sundews, and pitcher plants. Many are partially shielded by brush, grass, wild flowers and weeds.

I will point out in other parts of the book that sunlight is an important factor in helping plants attain their best colors. But sun *light* and sun *heat* are different. So is the drying-out process caused by overheating plants.

Your carnivorous plants will take on their best natural colors when given the natural sun that comes through those eastern and western exposure windows 6 to 10 hours each day. If you can't provide the natural light, you certainly can supplement it or even replace it artificially. That wasn't so true ten years ago.

Today, the Dura-Test fluorescent bulbs, sold as Vita-Lites, Naturescent lights, Sylvania's Gro-Lux, and several others are efficient. They have been developed to duplicate as close as possible the plant growth stimulating spectrum of light of the real sun. Even the newer Plant Lites, which fit into regular sockets, provide a substitute sun for plants in darker indoor areas at home or school.

Some experiments with lights are suggested in our chapter of study projects. In general, a twin-tube, 4-foot fluorescent fixture, suspended 18 inches over your plants, will produce satisfactory results indoors with any other light source. Run it 10 to 12 hours each day.

Under this completely artificial light source, flytraps turn their typical reddish color, sundews brighten, butterworts gain more yellow hues. The pitcher plants reveal their potential for coloration of reds, violets, and purples in their veins. Flowers especially show off more vividly under these new type supplemental lights or even when they are used as total sun substitutes.

Outside Growing

Many fanciers of carnivorous plants prefer to set their pets outside during warm weather. Some even plant complete bog gardens featuring carnivorous plants.

As long as you provide semi-shade so the sun does not bake your charges, the plants should do quite well. We have found that plants placed pot and all into the soil in a bog or moist woodland location will thrive all summer. You can, by using sufficient sphagnum moss, plant most sundews, pitchers, butterworts, and flytraps right into the ground. Bladderworts can be placed into a slow-moving pond or beneath a decorative fountain.

Keep in mind that humidity may be less outdoors on long hot days and in times of drought. Remember to water your plants periodically. A daily sprinkling doesn't hurt.

Come fall, only a few of the carnivorous plants can tolerate northern winters. The northern pitcher plant, some sundews and the northern butterwort varieties can overwinter.

Most other pitcher plants are natives originally to much more southern areas, the Carolinas and central California down to the Gulf coast. They must be returned to the cozy climate you provide them indoors if you want their pleasant company in years to come.

Once back indoors remember that the winter may be the time of needed rest for some of your plants. However, if you plan to keep them growing, the question always arises: what will they eat?

I realize you have a tidy home and flies just aren't available indoors. Maybe a few slip in each summer, but come winter they are all completely gone.

An outdoor planting of sundews, which require moist, sandy, acidic soil.

Don't kid yourself. Check an attic window any warm fall, even winter afternoon after a brief period of several sunny days. You're likely to find a fly or two that had hidden in some tiny crevice, nook, or cranny. He mistook that warm spell for a signal to wake up from his long winter's nap.

Use a little pill bottle or jar to scoop him up. Then just pop him into a terrarium or feed him to a deserving plant.

Another little trick works well. Bananas, apples, and other fruits are available all year long in stores. Just cut a few slices of banana or apple and leave them in your open terrarium. I assure you it is not spontaneous generation. There is no such thing. But even on the coldest days of winter, some fruit flies mysteriously may appear.

I've been told they hatch from eggs laid in banana skins and in turn lay more eggs in well-ripened fruit around the house. In schools, of course, you probably have access to those handy little fruit flies for science study.

We've talked to carnivorous cultivators who have fed meal worm bits to plants they thought deserved a winter treat. Meal worms and other natural insect foods are readily available at pet and tropical fish stores.

As you apply some of these cultural tips to your plants, you'll get the feel for tending them more helpfully week by week. They certainly deserve your best efforts. What other plants do you know that perform such amazing, inspired feats?

Venus Flytraps

Now you see it, now you don't. That accurately describes the way the world's most amazing plant lures, catches, and eats live insects.

One instant the unsuspecting fly or moth is hovering near the poised, open Venus flytrap. Sweet aromas secreted by the trap lead the unwary insect nearer. It lands while you watch and touches the tiny trigger hairs, barely visible on the inside of the cocked trap. That's all it takes. If you blink you can miss the action.

This remarkable plant can and does snap shut in a fraction of a sec-

Close-up, side view of a cocked flytrap leaf.

Just a tickle of its trigger hairs causes a trap to snap.

ond. We've timed a mature trap closing in $\frac{1}{20}$ of one second. The insect seldom has a chance. Equally remarkable is the power of the plant to seize and hold the victim.

Insects, as you probably know, have great strength. Compared to man and higher animals, insects have astonishing strength. Tiny beetles burrow through hardwood. The smallest, nearly invisible fly "teeth" can penetrate horsehide. Yet flies, moths, beetles, and many other types of insects are no match for the grip of the flytrap when it is hungry.

The Venus flytrap rightly deserves the description given it by famed naturalist Charles Darwin as "the most wonderful plant in the world." Over the centuries, this remarkable insect-eating plant has not only captured insects but the attention of some of the more astute observers in the entire field of botany and biology.

The first written mention of the Venus flytrap dates to about 1765 when John Bartram discovered the marvelous plants while visiting the American colonies. He collected specimens and sent them off to England for further study.

John Ellis, a leading botanist of that day, is credited with describing and naming the plant. Through the years many noted botanists, naturalists, and scientists in various related fields have conducted studies of this particular plant.

Carl Linnaeus, the acknowledged father of modern biology, called the flytrap a "miracle of nature." In correspondence with botanist John Ellis in 1768, Linnaeus noted, "though I have doubtless seen and examined no small number of plants, I must confess I never met with so wonderful a phenomenon."

As early as 1873, Sir John Scott Bourdon-Sanderson in England first discovered electromotive properties in the leaf of dionaea, using specimens that had been sent to the Kew Royal Botanical Gardens. According to him, "There is a definite electrical discharge that occurs in the Venus flytrap when the inner surface of the trap is stimulated. This action potential runs a course characteristic of animal nerve reaction." That was the first of many studies which verified the fact that this strange plant does indeed have what in plants can be compared to an actual nervous system.

Much later, in this century, a North Carolina physicist, Dr. Otto Stuhlman, also studied this electrical property of the flytraps. References to his works are included in the bibliography for those of you who wish to pursue this aspect of the plant's behavior.

Dr. Stuhlman traced the size of electrical charges that flash through the plant. He reported that the forces holding the trap open can be attributed to the action of an internal hydrostatic pressure. This turgor pressure causes the rapid closing of the trap. Measurements in his re-

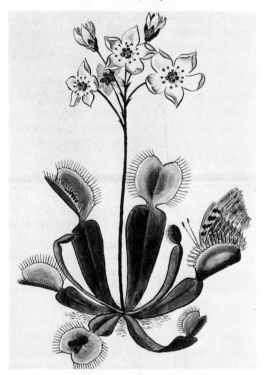

This artist's rendition of the Venus flytrap shows a typical arrangement of traps, from those just forming to those trapping or eating insects. The flowers are borne on a slender stalk, usually in May or June.

search reveal that the traps snap shut at the rate of 13 feet per second, closing in less than a half second. His studies are based on repeated tests which verify his and other earlier studies.

The phenomenon of the flytrap has provoked continued studies by scientists as well as students. One of the more remarkable reports we have seen recently was written by a tenth grade student in the Randolph School in Huntsville, Alabama. Miss Linda Donnelly carefully researched the earlier studies by a wide range of specialists in various scientific fields. We are indebted to her for permission to print some of her activities and conclusions.

"The scientific problem of my project," she reports, "was to research and investigate *Dionaea muscipula* and to find out if it were possible to measure the electrical resistance of such a plant during the digestion of an insect."

The first phase of her project was composed of preliminary research. A budding scientist, she dug deeply into previous works and writings about the plants. The second paper she wrote was the culmination of two years' work on the plant. That's dedication.

In her work Miss Donnelly used an "emotion meter," a transistorized electronic device from the Edmund Scientific Company of Barrington, New Jersey. The meter works by measuring the quantity of electrical resistance produced by the subject and registering the amount on a scale. It operates much like a lie detector, the polygraph which measures the electrical conductivity of the skin.

She secured the two wires to the midrib of the trap, that area commonly called the hinge of the trap. Adjusting the meter as prescribed in its operating instructions, she fed the plant a fly while stimulating the tactile trigger hairs until the trap began to close. She repeated the tests and took readings at intervals after the initial stimulation.

Chart after chart, which she carefully documented, revealed the same phenomenon reported by more learned scientists.

"I concluded," she reports, "that the results tend to verify that there is, indeed, a significant quantity of electrical resistance produced by a Venus flytrap during the original stimulus and actual digestion of an insect." She also added to the knowledge about this amazing plant.

"They also seem to indicate that a variation in such a response could be directly related to enzyme secretions and similar activity in the plant during early digestion. This hypothesis may be drawn from the fact that for long intervals there were lapses in the movement of the needle of the meter which were followed by periodic resumption of activity."

We trust she received a deserved A+ for her efforts. Her work demonstrates what can be done with imagination and dedication by others in exploring the mysteries of these carnivorous plants.

There are hundreds of plants that have developed remarkable insect-catching ability. Some wrap sticky tentacles around their victims; others drown their would-be meals in soupy broth.

But only the Venus flytrap has the astonishing ability to snap shut in less than half a second, often faster, then steadily tighten its grip until the insect is totally engulfed within the trap. Not only does the flytrap perform these fascinating feats, it thrives on its ability. This peculiar plant makes good use of the insects which it traps. They become, in effect, the needed nutrients for the plant.

Although other carnivorous plants exist, few demonstrate the rapid movement with which the flytrap catches dinner. While other types are found throughout the world, the Venus flytrap is only found in its original natural habitat within a hundred miles of Wilmington, North Carolina.

Almost everything about the flytrap stands in a class by itself. Perhaps that is what so impressed Charles Darwin. The many strange abilities displayed by this plant certainly have made it the best known and most popular of the carnivorous plants.

The flytrap is scientifically classified as *Dionaea muscipula*, and the term is really quite appropriate. At least it translates reasonably well. Dione is the mythological mother of Venus, the goddess of love. *Muscipula* is from the family name of the house fly, *Mussidae*. The *pula* part is the diminutive form, my Latin linguists tell me.

Consequently, the loose translation revolves around love for flies, perhaps. But then, who knows why and how our botanists arrive at the terms they use.

There are various theories why the Venus flytrap exists wild only in that certain area of the Cape Fear River's mouth. One holds that the particular area is favored by a unique microclimate and soil combination. The weather is favorable most of the year for plant growth, except for the few needed months of winter dormancy or rest. And the acid, silver sand soil has its own unique properties which are somewhat different from other areas and soil types.

The boggy areas in which the plant was originally found and the sandy soil are noticeably deficient in nitrogen and other nutrients usually required by most plants. Perhaps the flytrap's development was natural selection, adapting to the need for nutrients by devising its strange insect-digesting ability. We know it gets nitrogen from the insect's bodily protein.

Another theory, perhaps a bit far out, has recurred at times. That theory holds that the entire area of the Cape Fear River mouth was once either a giant meteor crater or was bombarded by a meteorite shower eons ago when the world was young. The meteorite pieces have been found, so that part is true.

There are almost as many theories as there are strange flytrap habits. Personally, I subscribe to the microclimate thinking. But then, perhaps the force of the meteorites had something basic to do with bringing up deep subsoil or even mutating plants that existed when the balls of fire struck the earth.

Whatever explanation you prefer, the fact remains: Venus flytraps are found naturally nowhere else on earth. However, arguments that they cannot be naturalized elsewhere aren't exactly true. They can be and have been in a few cases. For example, botanists and amateur horticulturists have succeeded in starting beds in South Jersey's pine barrens, around some bogs in Wisconsin, and also near the glades of Florida.

Studies on the Venus flytrap are being conducted now, as you read this book. Perhaps you too will join the legions who have and are still probing for the unlocked secrets of this plant. In that regard, from many reports received through the years and our own efforts as well, you'll find an entire chapter with suggested study projects on the various plants.

At this point, perhaps, we should review the plant itself, its parts, and some basics of how it works from what has been learned to date.

The flytrap begins like most plants from seed. More likely, any you obtain from specialists, garden centers, science supply firms, or other sources will be sold as bulbs.

Although most people believe a large bulb will produce the best and biggest plants, that is not necessarily true. We'll explain why later on.

The bulb, like most bulbs, is composed of layers or scales. They surround the central growing tip which is the flower stalk. It lies dormant until the proper time of year. Then, within a few short weeks it's up and blooming, pollinating and being pollinated.

The average-size bulb ranges between the size of a kidney bean and an acorn. However, even smaller bulbs the size of peas produce satisfactory plants. Larger bulbs cost more, but grow at about the same rate as the medium or smaller sizes.

Bulbs that have been preserved in cold storage have, in effect, been thrown into an artificial dormancy. When removed and planted they seem to sprout more quickly. Unless of course, they have been kept longer than a year or frozen in the process. Normal storage temperature is 36° to 40° F. Bulbs should not be frozen, since that can break down their cell structure.

The larger bulbs, when available, may have side bulbs attached. That's one other method of reproduction. As the plants grow over the years the bulbs expand. This usually occurs lengthwise, rather than just larger in diameter.

That means it is easier to simply snap the larger bulbs in half or thirds. Be careful to preserve some of those small, dark roots on each portion. In this way you can obtain several bulbs from one large or jumbo parent bulb. That, of course, means more plants for you and your carnivorous cultivating friends.

Flytrap bulbs, like most carnivorous plants, thrive in a growing medium composed primarily of sphagnum moss. You can use combinations of peat moss and vermiculite with a bit of perlite mixed in for good measure. A combination of acid soil and silver sand, half and half, covered by a layer of sphagnum moss also works well as a growing medium.

What type sphagnum moss? Whether you have access to living sphagnum near lakes, streams, or ponds, even drainage areas in your area, or use the store-bought dried material, either serves equally well.

One flytrap plant fits easily in a 2-inch pot.

The living moss greens up and grows, which can be a problem. It can smother or overcrowd the plants unless tamped back in place periodically.

See the chapter cn culture and care for more growing tips. For the time being, whether you plan a pot or small terrarium planting, you're well advised to insist on sphagnum moss for the growing medium. It has several advantages. It holds moisture well, yet does not pack or mat down. That means roots can penetrate more easily. So can air. And, plant roots, like leaves, need a certain amount of air circulation to perform properly, even those small roots of the strange flytrap.

Most bulbs arrive, whether in store package or mail order, with tiny dark roots. Some have been trimmed. These too will sprout and grow. The rounded portion of the bulb is the base. The part with the leaves is the top. That may sound simple, but usually the leaves have been trimmed off too and the bulb packed or shipped with just the stubs of old leaves left.

Place the bulb with the top just at the surface of the sphagnum moss. Firm the moss around each bulb.

Water well. But please don't use city tap water that is loaded with chlorine and possibly fluoride. Let that water stand a day or two.

Flytraps and other carnivorous plants can't stand chemicals in the water. That goes for hard water with high mineral content too.

You're better advised to use rain water, well or lake water, providing the stream or lake is not polluted. Distilled water can be used, but good old rain, well, or stream water does have some minor amounts of nutrients in it which seem to benefit the plants.

It is best to cover the plant with a plastic cup or dome, unless you prefer a terrarium planting. As it sprouts, the plant enjoys lots of humidity. That's true for most sprouting bulbs and plants.

Within several weeks the leaves will begin to form the first tiny traps. You can compare your plant's progress with the photographs of emerging traps in this chapter.

As the bulb develops its leaves, you'll see them in various stages. Some are just emerging, others are beginning to open, while the first up will have begun to cock open to begin their insect-attracting work. Depending on the warmth (65° to 75° F. is good) and humidity (which should be constant and over 50 per cent if possible), the bulbs will produce from 4 to 12 traps within 6 to 10 weeks from planting time.

Occasionally a few traps turn brown or emerge with tall spindly growth. It happens. The problem may be too much heat, which can occur when you place the planter in a full sun window and soak too much heat into the growing unit.

Another problem can be too much light, or conversely too little so the plant struggles to get its fair share of sun. Usually, better-looking traps form within the initial 8 to 10 weeks of growth.

All traps form at the end of each leaf. During certain times of the year, usually early spring in natural habitats or when using bulbs from refrigerated storage, the first leaves will have flared ends with small traps. Don't worry. That's a natural situation.

Larger traps develop on the leaves that appear a bit later. Of course, as the plant begins to dine in style and flourish from that nourishment, the traps just naturally become larger as new leaves continue to sprout from the bulbs.

As the trap forms you'll see the beginnings of the fingers which line the edge of it. Eventually these fingers, technically called cilia, will be almost a quarter inch long on the larger traps.

Once the trap begins to open fully you'll be able to see the tiny trigger hairs. There are usually three per lobe, the lobe being one half of the trap.

When you look down at a trap it seems to be hinged in the middle. In effect that midrib performs a vital function. It actually performs like a hinge. When fully open, the cocked trap resembles your hands if you

Side views of the flytrap.

placed the heels of each palm together and opened your hands, with fingers extended and arched.

The open trap seems quite simple, until you think about the strange functions all those little cells can perform. If you touch *one* of the trigger hairs, nothing usually happens. If raindrops sprinkle down on the traps, nothing happens. You can prove that when you water the plants.

Look again at those fingers. They are spaced from ⅛ to ¼ inch apart. That means if a small insect accidently triggers the trap it can most likely escape. In most cases, the smaller gnats and fruit flies only touch *one* of the triggers. That seems to be nature's plan. Those small insects just don't have sufficient food to be worth the plant's efforts to snap shut on them.

However, here's where the fun begins. Take a pencil or small twig. Gently touch *two* of these nearly invisible triggers. Or, one hair twice. When you do, the primitive nervous system gets the message: food at hand! Try a toothpick, held lightly. The trap will snap it right out of your fingers and hold it.

When a larger insect, a moth, butterfly, ant, beetle, or fly taps the

This series demonstrates the triggering action of the flytrap. 1) Cocked open. 2) When stimulated by an insect. 3) Final closure to entrap the victim completely.

Even large moths landing on the enticing surface of the Venus flytrap are potential meals for this plant. Note other traps still waiting, one with the skeleton of a digested insect in it and another fully closed, digesting.

trigger the reaction is the same. The trap snaps, fingers interlocking to prevent the insect's escape. Now watch the action when a fly or small insect visits the cocked trap. Once the initial snap occurs, the insect begins to struggle. After all, it must realize there's something terribly wrong when a plant snaps shut on it. The more the insect struggles, the more it stimulates the trigger hairs, and the enzyme-secreting glands along the inner lobes. The greater the struggle, the tighter the trap closes, steadily, bit by bit, until it is fully closed. Then, as if by some primitive instinct, the fingers flex outward.

But it is too late for the victim. The trap has sealed its meal in an airtight pocket. More enzymes pour out to begin the digestive process, drowning and digesting the insect.

Some scientists also point to the possibility, more a probability, that those enzymes and other secretions also have a tranquilizing or anestheticlike effect on the insect. Perhaps it is, they theorize, much like the bite of a spider which paralyzes its victim until it can be laced over the webs.

For many years, credit for the digestive process was attributed to action of the enzymes. That seemed logical. After all, no other action seemed to occur in the thousands upon thousands of flytraps we had grown under a wide range of conditions.

Often, when a seemingly logical answer has been provided, we drop the matter and accept that conclusion. Some people don't. They want to get the truth themselves, such as why the flytrap is so restricted in its habitat.

A plausible explanation has been set down by Dr. B. W. Well of North Carolina State University. His theory is that the plant was rendered almost extinct by the cold glacial climate many thousands of years ago. A few plants managed to survive on the Cape Fear peninsula. Since then it has had time to migrate less than a hundred miles from that point.

Other didn't fully believe that only enzymes accounted for the flytrap's digestive process. Although we have not seen copies of the research done at the University of Southern California's Department of Bacteriology, it was, I understand, aimed at discovering the part that bacteria play in the digestive process for insectivorous plants. In one report by Thomas Emmel on that subject, he notes that bacterial action does play a role in digestion in consort with the enzymatic action within the closed trap.

More research should be conducted along these lines. There's no telling in advance what may be revealed. Remember that streptomycin and penicillin were discovered from the most unlikely substances, the first from soil, the latter from mold. Perhaps new efforts to probe the secrets of the flytrap in all aspects of its mysterious processes might lead to similar valuable discoveries. Stranger things have happened.

Once the insect has been trapped and the digestion completed, the trap, from some internal secret signal, will reopen. It must know that dinner is done. Usually a trap can digest an insect in a week to ten days. When it reopens, you'll only find a husk, what amounts to the inedible part of the insect. That makes sense. After all, you leave the bones when you eat a steak or chop, right? Rain, wind, or just watering the plants will remove the few remains. That same trap may eat several

After the trap reopens, only the husk of the victim is left, to be blown away by the wind or washed away with the rain.

additional insects before it begins to dry, turn brown, and die. That process is to be expected. Even if a trap doesn't ever catch an insect, you'll find that the older traps die periodically, almost in programmed sequence. Happily, others emerge from the bulb to continue the catching, eating, and nourishing process for the plant.

A strong, mature plant may have 6 to 15 traps of various sizes and stages of maturity. We have seen plants still thriving after several years of continual growth. They have expanded to 20 or 30 traps. Some of this can be attributed to the growth of the bulb, which elongates and can be divided into several bulbs. In effect, there are what amounts to several bulbs producing that profusion of traps.

Many people, from students to advanced botanists, have tried repeatedly to produce plants with larger traps. They have applied plant growth hormones of various types, bombarded seeds and bulbs with radioactivity to achieve mutations. To date, no one has been successful in his quest for a giant trap. We also have tried to achieve the same goal in many ways, from special cultural efforts, including force feeding of plump live flies, to variations in planting mix, fertilizer, organic soil nutrients, and lighting combinations. We also tested gibberellic acid and similar growth stimulators.

The best results came when we placed plants in a large aquarium, as a terrarium. Then, we maintained the humidity over 60 per cent, provided Vita-Lites 16 hours each day, and introduced dozens of flies periodically. They were nice and healthy, too, direct from a neighboring dairy farm. After a year the traps were close to 2 inches long. The light helped turn the plants the typical bright red to darker scarlet you will find when plants have full daily sunlight.

You may be wondering why your traps aren't a bright red. Although many plants will produce a profusion of traps, they may not produce the desired bright red color on the inside of the lobes. They do still perform their insect-catching process. Adequate light is the answer for the desired red coloration. We have used the Dura Test Company's Vita-Lites and Natur-escent lights, both fluorescent types.

Over our experimental and propagation beds as well, we suspend 4-foot, twin-tube fluorescent fixtures. Sylvania Gro-Lux tubes work well but must be closer to the plants. We hang our fixtures 10 inches above the traps initially. When bulbs have sprouted, and first traps begun to form, we raise the lights to 18 and as high as 24 inches above the plants. In all cases, we have achieved bright red traps on almost all the plants. They also seem healthier, but that may be our imagination since they may just appear more attractive.

Can you keep the plants alive and thriving year after year? That's a typical question. Yes, you can. Some won't continue and may even die

The inner lobes of the traps turn bright red when the plant is given bright sunlight.

back. In fact, many people report that their plants live and perform well for many months, then seem to brown and die. There's a logical reason for this. The flytrap is produced from a bulb. Like all bulb plants, they need a period of dormancy to rest. It is part of their natural cycle, similar to the cycle of tulips, daffodils, and other plants that grow from bulbs.

You can provide that resting period in several ways. First, cut back on the water you provide. Traps will dry and new leaves fail to sprout. Cut away old leaves and traps. Then, remove the bulbs, clean them, and place them in a refrigerator. Best holding temperature is 36° to 40° F.

After 8 to 10 weeks, remove the bulbs and replant them. You'll be surprised how quickly they resprout and begin their life cycle again. We've done this, using varying storage periods. Maximum storage life is about 18 months, but that may cost you some bulbs which soften and die in storage. This refrigeration can be compared to duplicating the usual winter rest for plants.

In spring the flytrap sends up its flower stalk which bears several lovely white blooms. (Courtesy Carolina Biological Supply Company)

Another way to rest bulbs is simply to reduce their water supply. Allow the bulb to dry. Then trim off the top and let it rest in place. Be sure to provide some moisture, since bulbs should not dehydrate completely. This process is a duplicate of summer dormancy in nature when dry weather or droughts would put the plants to rest. The summer dormancy can be misunderstood. Many people see their plant just naturally begin to turn brown and find no new leaves emerging. They assume the plant has died, so they dig it up and throw it out. That's a waste. In most cases, unless the plant has died from other causes such as excess heat, dryness, overwatering, or even harmful chemicals in the water, the bulb is most likely still alive. Don't discard it. Try refrigeration and a resting period. In many, many cases bulbs will snap back, alive as ever, when replanted.

Flytraps customarily bloom in May and June. If you have plants growing that haven't bloomed, take heart. Plants sprouted from bulbs kept in cold storage may have their cycle slightly altered the first year.

We've seen them bloom in just about every month. That's under-standable. They are just responding to their built-in signals and bloom-ing within the period they usually would send up flowers after awaken-ing from a winter's rest.

Flytrap flowers are surprisingly attractive for so "nasty" a plant. Flower stalks emerge from the center of the bulb, normally one stalk per bulb unless it is a larger one in process of dividing into several bulbs.

You may have just a few or up to ten flowers at the top of the 6- to 10-inch flower stalk. They range in color from white to pinkish white, as illustrated in the photos.

One by one they dry and seed pods form. Flytrap flowers are self-pollinating, so let the pods dry until they are dark brown to black. Then wait a bit longer until the seeds inside ripen and dry.

Although you can propagate flytraps in several ways, the most chal-lenging is from seeds. Cut off the flower stalks and place them on a piece of white typing paper or other white surface. Then, cut or snap the pods open. They're small. Take your time so you don't lose your seed crop. You can store them in a plastic sandwich bag in the refrig-erator until ready to plant them. Or, sow them immediately.

Prepare a seed bed of chopped sphagnum moss, not too coarse or too finely ground. Run a few handfuls of moss through a blender set on "chop" and you should obtain the right consistency. Place the sphag-num moss over a layer of regular sphagnum moss in a tray, bowl, or other container. Scatter the seeds lightly on the surface and then sprin-kle a fine layer of the ground moss over them. Moisten thoroughly, using a fine mist spray. Then, cover the tray or planter with a plastic hood and place it under the propagating Vita-Lites. Sunlight on a win-dowsill will be satisfactory too. Germination may be slow. One factor that may cause failure is drying out during this critical germination stage. Make certain your seedbed is moist at all times.

Within a month, or up to several months, seeds should begin to sprout. Look carefully. Emerging from the seeds, you will see the smallest traps believable, some less than ⅛ inch long. They may be tiny, and there may be dozens of them. But even these diminutive traps can catch those minute fruit flies to begin their long life cycle right be-fore your eyes.

If you prefer merely to enjoy the plants you grow from bulbs, you can let them flower. However, when the Venus flytrap flowers you'll notice that traps nearly disappear. The answer is simple, but often misun-derstood. The bulb is spending a great amount of its stored-up strength to flower and set seeds. To encourage more traps, simply cut off the flower stalk and new traps will begin emerging.

There are other easier ways to propagate this amusing plant. One of the most reliable is bulb division. After plants have thrived for several years, the bulb just naturally gets larger, but not necessarily in diameter the usual way. Instead it elongates horizontally. In effect, new bulbs form almost as side bulbs. These can be simply snapped apart and replanted to produce two or three, even up to five new plants from the largest older bulbs. When you do this, try to keep several roots on each division.

Another way, admittedly more difficult, is vegetative reproduction of bulb scales themselves. You can peel these scales from even medium-sized bulbs. Then immediately plant them in the same medium suggested for the seedbed. Although this method is not as successful as bulb division, and may not produce more than half the new plants you expect, it can be accomplished if you maintain the required high humidity and warmth. Some horticulturists have tried heating cables in the planting mix as you would for vegetable and flower propagation. It seems to help, but statistically does not seem that significant for increasing the quantity of new plants.

The final way to root new plants is similar to conventional leaf cuttings. Remove several leaves from a healthy, sturdy plant right down to the bulb. Be sure to leave a small portion of the whitish bulb on the leaf. Then, insert these leaves into moist sphagnum moss the same as used for seed starting. Keep these leaves sprinkled daily. If they dry out you put too much stress on them and they can't set new roots well.

A misting system, sold by home greenhouse firms and garden centers for mist-propagating plants, is helpful. It keeps the leaves well moistened so they can concentrate on rooting. Laying the leaves on the moss with tips buried works well too.

For those of you who want to grow and enjoy these plants, those are the basics, right through simple to somewhat more complicated propagation methods. A few more thoughts are in order now, before we continue into some fascinating technical studies and research for those of you who want to expand your growing horizons with these plants.

You can tickle the flytrap with finger, pen, toothpick, or other object and it will snap shut. When it opens again, you can make it snap again, and again. However, you'll notice that each snap is slower than the last. When you fool the trap, and it doesn't find an insect to digest, you weaken it a bit. Plant cells, just like all cells, wear out. New ones can form, but these unique plants need nourishment to generate new cells and rebuild the turgor pressure in former cells. We recommend from long experience that you control your urge to tickle the traps too often. They'll do better when they utilize their energy to catch real meals.

Traps *will* snap shut on a piece of hamburger. However, the quantity of protein in a chuck of hamburger is often just too much for the individual trap to digest. Remember, you don't want to kill your plant through misplaced kindness, do you? For experimental purposes or periodic demonstrations, go ahead, using the smallest sliver you can.

Although most of the growers who cultivate the Venus flytrap do so for the fun of watching it lure and catch insects, some scientists have really concentrated on learning more about the whys of the trap's strange mechanism.

One experiment focused on discovering what really makes the trap snap. In a series of experiments, scientists connected tiny electrodes to portions of mature traps. Wires ran to a laboratory oscilloscope. This device is utilized to record electrical impulses of direct or alternating current. Is it possible that a plant actually generates electricity? So it seems.

When all was ready, the trap was tripped. Surprise! Imagine the awe of the researchers when they actually recorded evidence of an electrical discharge. They saw the wave pulse right across the screen of the oscilloscope. In repeated tests electrical current was produced again and again by different traps.

In continued studies, these investigators and others who expanded on the original work learned that the flytrap indeed gives off a small, weak, but nevertheless real electrical discharge when its trigger hairs are touched. They have theorized, and it would seem logically so, that it is the electrical discharge which sends the signal to the cells, changing the water or turgor pressure in the cells and causing them to act in unison. The result is the amazingly rapid snapping action.

To perhaps make this more understandable, place your two hands with the heels together as you would to demonstrate the shape of an open flytrap. Extend your fingers so your hands resemble the cocked trap. Now, think! Then snap your hands together with fingers interlaced to simulate the rapid closing of the flytrap. What happened when you thought was: Electrical impulses ran from your brain to the muscle control center of your brain for your hands; the impulses ordered the muscles to close your hands, which they did. That, in perhaps oversimplified terms, is the same principle that applies and occurs when the flytrap snaps shut.

A strange plant indeed, with a proven primitive nervous system and capable of generating electrical current.

Other researchers are exploring different aspects of this plant's weird habits. What, they ask, are the properties of the insect attractant secreted by the traps? What is the chemical composition of this material?

They have good reason to ask. They theorize that if they can learn the chemical composition of that insect attractant and artificially re-create or synthesize it in the laboratory, they may be on the verge of a breakthrough in biological insect control.

Consider, for a moment, the advantages of having just such an effective insect attractant. You could place containers of it near farm fields and home gardens. Insects could be lured away from the valued crops into special containers. Once inside, they could be destroyed electrically or chemically. That process could do away with large-scale spraying of pesticides on food crops. Those scientists are not just dreaming of blue-sky discoveries. They have rather practical applications in mind for development once they discover the flytrap's alluring secret.

Undoubtedly there are other equally unusual experiments that have been or are presently being undertaken with this plant. It certainly has captured man's attention over the centuries. After considering all the plants of the world, interviewing hundreds of scientists and individuals on many horticultural subjects, I'm convinced that the flytrap retains its justified place in the plant kingdom.

It is today and probably will remain what Charles Darwin said it was, "the most wonderful plant in the world."

Sundews

Glistening redly in the sun, tentacles aquiver, sundews seem too attractive to be carnivorous plants. The Drosera family is a large one, with varieties found in all parts of the world.

Sundews, like flytraps, actively lure, catch, and eat live insects. In fact, some of the larger sundews are capable of snaring far more insects for their size than any other carnivorous plants. The tall southern filiformis, for example, has a remarkably hearty appetite. We have counted up to one hundred tiny insects securely glued to one of these 15-inch-tall plants. Among carnivorous plants, sundews are probably the greediest of them all.

Sundews range throughout the world. We have seen them and studied them on several continents in a score of countries. To the best of my knowledge, some varieties, and often several, grow wild in every state. Fortunately for the timid-hearted, there are no really giant sundews here, as exist in more exotic parts of the world. The hungriest of all the sundews, with leaves almost 2 feet long, do grow in Africa. There, with favorable humidity and long days of strong sun, sundews are capable of catching a variety of small animals that venture too close for their own good.

Sundews, all types, are considered active plants. True, they don't have the speed and agility of the notorious Venus flytrap. Still, they are well equipped with hundreds of sticky tentacles. Even the smallest varieties have ample sticky arms to wrap around their insect victims.

This family of fascinating plants also captured the attention of Charles Darwin. Since his original studies, which were extensive, many others, from amateur naturalists to accomplished scientists, have probed the secrets of the sundews. During my twenty-plus years with carnivorous plants, I've found these botanical oddities a constant source of wonder. Every time you think you have solved a riddle, the plants perplex you again.

The tall sundew, *Drosera filiformis*, sprouts from bulbs
to form a dense tangle of sticky leaves.

Most sundews are fairly similar to each other. Some are so closely related that it is extremely difficult to tell them apart. Of course, natural
crossing further blurs the lines among relatives in this plant field.

Others are strikingly different, as you will learn in this chapter. One
variety, which we obtained less than a year ago, is astounding. It's a giant version of our familiar *D. filiformis*, which generates several dozen
15-inch tall arms. Each is covered by hundreds of sticky hairs. We
planted this specimen in one pot and carefully tended to its needs.
Then, we introduced a culture of drosophila, the common fruit fly, into
the surroundings. Within one week it had eaten my entire colony of
fruit flies. Every leaf was covered. At that rate, it's as bad as buying
meat for a growing family. Who can afford a colony of fruit flies each
week per plant? And now, we have several dozen growing merrily away.

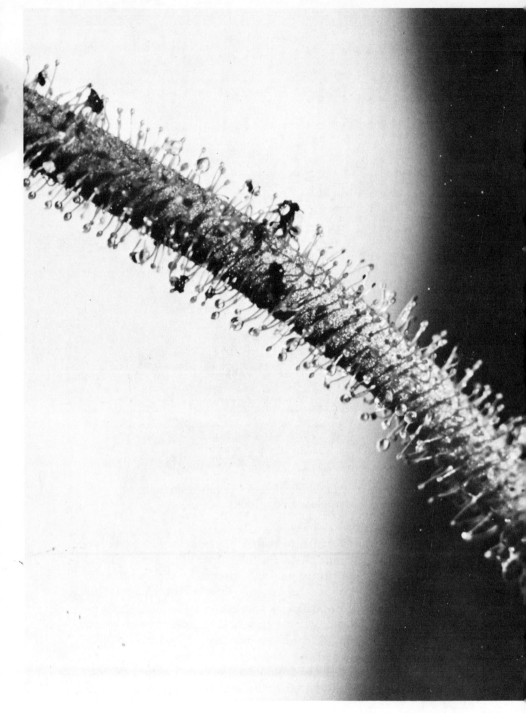

Tall sundews have hearty appetites. For their size, they catch more insects than any other carnivorous plant.

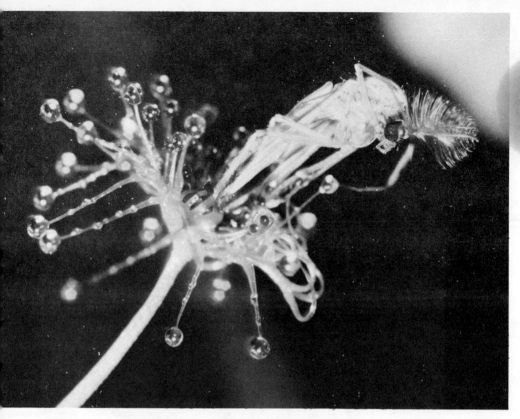

Even larger insects are held firm by the sundew.

Worse yet, what will happen if I cater to their hearty appetite? But then, I've always been searching for that legendary giant among the carnivorous kingdom.

The most common and popular sundews to date have been the smaller types which grow so well in terrariums, pots, and mini-planters. Even among these smaller ones you have a wide choice of varieties. Before discussing the distinctions, let's review the techniques and talents of these sticky, tricky plants.

Generally, most sundews grow from seeds dropped or popped out by the preceding year's plants. As they sprout, arms radiate from the center in a rosette pattern. This pattern holds true for all the sundews.

At the end of each arm, the small leaves form. The most common, *Drosera rotundifolia*, features a round leaf. Others are pear shape, elongated, oblate, or spatulate. To make those terms more understandable, we've included sketches with the terms further along.

Rotundifolia leaves may be as small as ⅛ inch or as large as ½ inch. You'll find the largest plants may have one to two dozen leaves per plant. On each leaf are dozens of tiny, sticky hairs, more appropriately

Typical growth of the rotundifolia sundew. (Courtesy Carolina Biological Supply Company)

known as tentacles. At the ends of these hairs are small black spots, the glands which secrete the so-called dew. This dew is in reality the sticky substance which glues the unwitting insects tightly to the plant until other tentacles can add their strength to ensnare the plant's meal.

When an insect is attracted by the secretions from the aroma or attractant glands, it may land ever so lightly on the glistening leaf. Oops —one foot gets stuck. It flutters its wings; they become stuck. The more it struggles the more the sundew is stimulated to secrete more fluid as well as to begin releasing its digestive juices. Foot by foot, wing by wing, the insects are engulfed and trapped by these tentacles on each leaf.

The sticky glands and their secretions are clearly shown in this group of tall sundews.

It doesn't really matter whether the sundew leaf is spatulate or pear shape, oval or round. They all have similar tentacles and glands which respond in an identical way to catch their dinners. These lovely little plants glisten beautifully at dawn, as though the morning sun was sparkling on the dew. Some dew. If you take a finger and apply it to a spar-

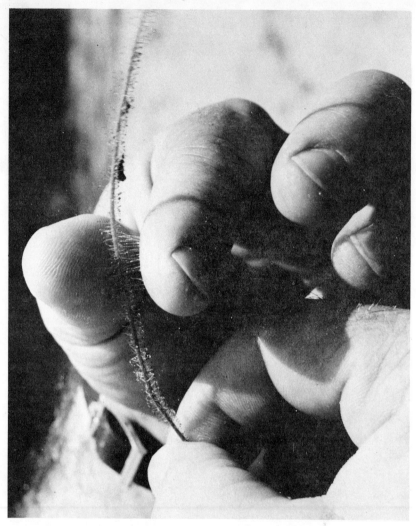

Note the holding power of the tall sundew's mucilage.

kling sundew, you'll feel and see the tacky substance stick to your finger. Lift it away and strands of the substance stretch out as though attempting to hold even your finger to the leaf.

Foreign sundews have similar habits, although their appearance may be quite different. We've included some photos to demonstrate the range of appearances in the exotic Drosera family. All, of course, thrive on living diets of insects all the way up to small rodents, animals, and birds.

A tall sundew in bloom.

There are taller types, some with long stems on which the sticky leaves waft in the breeze; there are other varieties in which every portion of the leaf can catch the prey. The filiformis, of which several varieties are native to the United States, possess the strongest appetites of our domestic sundews. Unlike the typical sundew, these sprout from small bulbs. In areas of southern New Jersey and the mid-Atlantic states you can find them glistening along roadside ditches. They prefer the usual moist, boggy conditions, but also thrive in wet, sandy locations.

These bulbs can be kept in refrigerated storage, unlike the usual sundew varieties which grow from slender roots. Those must be kept alive or they'll die from the cold. In spring, a few filiformis leaves begin to

(Left) The fly has just been captured. (Right) Note the flexing of the tentacles as this sundew leaf digests its meal. (Courtesy Carolina Biological Supply Company)

sprout. Following these juvenile leaves, the larger ones appear. They unfurl much like typical growth of ferns.

Filiformis is native along the eastern coast and in some midwestern and western states. It grows 6 to 10 inches tall. Its giant, close relative, with magenta rather than lavender flowers, matures up to 15 or 18 inches high. It has the same growth pattern, many leaves unfurling from a somewhat larger bulb.

The conventional tall sundew arises from a bulb the size of a pea. The larger relative emerges from a bulb the size of a hickory nut or acorn.

Both these filiformis varieties have thousands of sticky hairs along each long, slender leaf. We have tested them under artificial light, attempting to achieve greater coloration in the leaves. Unfortunately, they remain the same yellowish to golden color as they do in nature. In their natural habitat these sundews are at best pinkish yellow to a golden color. The taller version seldom seems to change from greenish yellow or light yellow.

Because both types have such sticky tentacles, you can actually place live flies and small moths against the leaves and stick them fast. In feeding trials, when we and others have deliberately attached dozens of insects to these plants, they responded favorably. More leaves unfurled from bulbs until we produced a veritable bush effect from one plant. No doubt its diet agreed with it.

Intermedia sundews, common along the edges of wet ditches and bogs, are thirstier than their relatives. Most sundews thrive when grown in a mixture of sand and sphagnum moss. Some actually grow in sphagnum alone. When it overgrows them, they flower, set seeds and replant themselves, layer by layer, in the bog.

This cluster of sundew plants in a glass bowl shows a gnat's-eye view of the hundreds of sticky tentacles.

Intermedias, on the other hand, require planting in containers with constant water for their roots. The roots are slender, delicate-appearing. They probe into the mud and sand to gain a roothold. During rains that fill these roadside ditches, intermedia can be completely swamped. That doesn't seem to bother them. As the water level subsides, there they are, bold and sparkling, ready to snatch another insect who alights on any leaf.

The rotundifolia, intermedia, and filiformis are probably the most common sundews available through science firms and plant specialists. Several carnivorous plant growers offer other varieties so you can obtain a wider range of these curiosities of nature.

Pitcher Plants

Pitcher plants may appear as lovely, colorful, somewhat curious specimens, but they are as lethal to insects as the more aggressive Venus flytraps or the tentacled sundews. It is difficult to determine whether pitcher plants are really less advanced in their insect-trapping methods, or just more devious in the way they catch their victims.

No doubt they have devised equally effective ways to lure those unsuspecting insects to their patiently waiting, yawning mouths. When you have watched as many pitcher plants as we have, in fields, nurseries, greenhouses, and in controlled experiments for years, you become accustomed to their ability to attract a vast hoard of insects. They just keep flying by and disappearing into those hungry, open mouths.

The open mouth is common to the pitcher plants. So is the digestive process. But there the similarity ends. You can tell they are from the same family easily enough. The differences they display in coloration, shape, form, and size are what make pitcher plants such intriguing subjects for study.

Pitcher plants, like sundews, are found in scattered parts of the world. The United States is honored with a fine range of these particular carnivorous plants. Since all the varieties that have been found native to America are easily cultivated, we'll focus first on these.

Perhaps the most popular is the so-called purple pitcher plant. It is often called the northern pitcher plant, but its southern cousin is closely related. Given adequate sunlight, both turn dark red to almost crimson purple.

Looking at one for the first time, you might be deceived. It seems harmless enough—it is to you. It appears to be just a radiating cluster of hollow stems, a bit bulbous on the bottom, in which some rain has collected.

How deceptive these carnivorous plants can be! In reality, the *Sarracenia purpurea gibbosa* and its southern cousin, *S. purpurea venosa*,

A typical purple pitcher plant, here growing in a clay pot. (Courtesy Burgess Seed & Plant Company)

are passively poised for their prey. Unlike the flytrap and the many sundews, the pitcher plants have no way of snapping out or wrapping sticky fingers around their prey. So wait they must. During the centuries these plants lay waiting, they perfected an ingenious method to catch the insects they need to supplement the nutrients in the leached-out, boggy soils in which they live. Each pitcher sprouts from the central rhizome, supported by sturdy roots which hold it in the moist, sandy, and moss-covered soil.

Around the mouth of each individual pitcher are hundreds of tiny spines. These ⅛-inch spines, even ¼-inch long on larger plants, line the mouth and throat of the pitchers. You might think they would point up toward the sun. Instead they point down. There's method in this scheme of things.

An insect's-eye view of the spine-lined mouth of the purple
pitcher plant.

As insects are attracted to the plants by secretions from insect attractant glands, the moths and crickets, flies and beetles enter cautiously.
Inch by inch they follow the trail of spines inside the pitcher. They
seek what seems to them to lie ahead. If any realize the danger, they
may turn to flee. That's where the pitcher plant was smart as it evolved
over the years. Those down-pointed spines are slightly flexible. Insects
may push them up a bit, but soon enough the spines push back.

Despite its efforts, the insect is flipped backward. It may try again as
some will do. No doubt there are the stronger ones that do escape.
More likely, they either follow the trail of spines right into the soup or
are flipped backward into it when trying to escape. The soup we speak
of is the liquid brought up into the pitcher by the roots. Enzymes
secreted by glands inside the pitcher have digestive properties. Rain

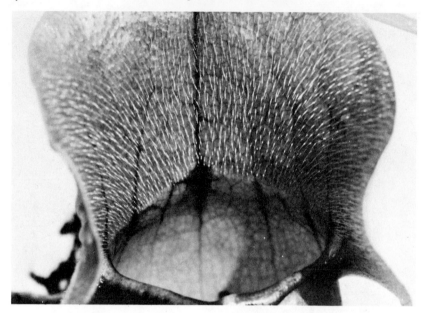

Inside the spiny mouth of the purple pitcher plant.

may dilute them temporarily, but the extra liquid serves to drown the insects, too. Eventually the soft portions of the insects are broken down by the digestive fluids of the plant, to become its meal.

As the plant prospers on the nourishment provided by its insect diet, it may also grow sufficiently large to capture small frogs, a stray caterpillar or two, even a salamander or other more exotic fare.

During its growth cycle new pitchers emerge from the rhizome periodically, in 6- to 10-week intervals, depending on the season and the plant's luck in catching food. Older pitchers turn brown, wither, and eventually decay. The new ones keep sprouting all season until the chill of fall, when all foliage dies back.

Although there are slight differences between the northern and southern versions of *Sarracenia purpurea*, both bear almost identical, lovely, red velvety flowers. Flower stalks spring up from March to June, depending on how the season has progressed in the area in which the plants are growing. We've seen the southern variety bloom in early March in Georgia. The Maine and Canadian pitchers usually blossom by June.

Checking our historic sources, we find that there have been references to pitcher plants by the earliest settlers. They probably learned about these plants from the Indians, since occasional records indicate

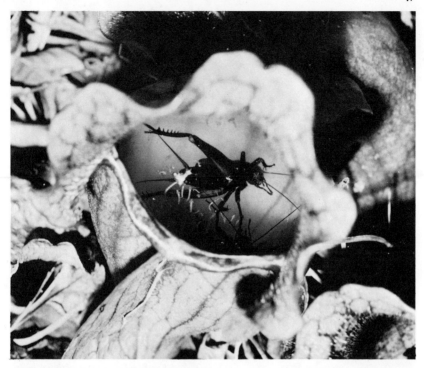

The captured prey—here, a cricket—inside a purple pitcher leaf. (Courtesy Carolina Biological Supply Company)

that Indians used some portions of the plants for folk medicine in early Colonial days. In records from the 1600s we find that a Canadian, Dr. Sarrazin, collected plants and sent them back to Europe for study. These early investigations by botanists led to the naming of the genus Sarracenia in honor of their discoverer. In all likelihood the first pitcher plants identified were purpureas, the northern pitcher plant, since that is the main one found native in North America.

It was known in the early 1800s that insects were attracted to the tall pitcher plant, *Sarracenia flava*, and seemed to be paralyzed by some unknown substance in that plant. Modern scientists have hypothesized that perhaps the aroma of the flowers or pitchers themselves not only lured the insects, but also acted in some way to paralyze them so they would fall into the digestive fluids inside the pitchers.

Dr. Howard Miles of Mississippi State University is intently concerned with chemical research. His team of scientists is analyzing various plants for anti-tumor activity. Some plants, including the pitcher plant, indicate promising activity, he reports. Betulinol and lupeol have

The red, velvety flowers of the purple pitcher tower over the plant.

been identified as having such capacity. Pursuing these findings, Dr. Miles and his associates, including Naresh Mody, Rodger Henson, and Paul Hedin, steam-distilled nearly 200 pounds of the *S. flava* plants to obtain sufficient amounts of the chemical for continued research. Dr. Miles did indeed identify the insect-paralyzing agent as an amine known as coniine. It is, he notes, one of the volatile alkaloids of the

Typical growing arrangement of the purple pitcher plant.

toxic hemlock plant. If the paralyzing agent proves as effective as initial studies indicated when tested against fire ants, now migrating across the southern states, new breakthroughs in insect control may be achieved.

When you examine a purpurea pitcher plant closely, you'll find that it and other related pitcher plants include a rhizome with fibrous roots; some a few inches, others 6 to 10 inches long. All these plants send up their pitchers in a more or less rosette pattern. This pattern is more obvious in those that lie close to the ground. The pitchers themselves are hollow. In the case of purpurea they display a large bulbous area in which the digestive fluids blend with rains.

From seed, tiny cotyledons, the first primitive leaves, appear. They're almost too small to spot in wild marsh conditions. In seedbeds, trays, or pots you can see them emerge, followed shortly after by the more typi-

cal hollow pitchers. If you are fortunate to find some in your locale, examine the surrounding area closely. Often you can find a scattering of
young plants around the parent, much like baby chicks surrounding
their mother hen.

During the insect-catching season these adult leaves perform their
functions well. By fall they are often replaced or intermingled with the
so-called winter leaf. It may have a mouth but seldom has the fully
open hollow pitcher. These leaves look more like wings with a mouth at
the end. When cultivated indoors, you'll most likely be successful in
maintaining a continuing emergence and growth of the desired hollow
summer leaves.

Look inside these hollow insect-catching devices. In fact, remove one
and slit it with a knife to better view the secrets that lie within. Across
the spine-lined mouth, which is the attraction zone, you may be able to
detect the nectar glands. They're responsible for luring the plant's potential meals. There are fewer hairs or spines in the lower "throat" area.
Below that is a more waxy surface. I suspect that once the insect passes
the spines it is supposed to slide the rest of the way into the "soup."

By careful examination you may detect the glands which secrete the
digestive fluids and enzymes in the lower portions of the pitcher. Once
you have come to this point, the area to which the liquid usually finds
its level, you'll begin to see the insects in the various stages of decom-
position and digestion. Finally, the pitcher narrows to the short
stemlike portion which connects the pitcher to the rhizome.

This cross section of a pitcher plant leaf reveals the partly digested prey
and fly maggots too.

Although many insects become the nourishment for these plants, others do escape. There is an interesting sidelight here: some insects purposely cohabit with the plant and join it in its catching fun.

Parrot Pitcher Plants

The parrot pitcher plant is well named. It also forms a rosette growth pattern, with pitchers radiating from the central rhizome. Although it is typically smaller and less conspicuous than the northern pitcher plant, we have grown some which have measured 15 to 18 inches diameter with individual pitchers 8 inches long.

This particular species is a sun-loving plant native to the bogs and savannas of the southland. It grows well enough in domesticity as a house plant, in terrariums, or in outdoor bogs during summer in north-

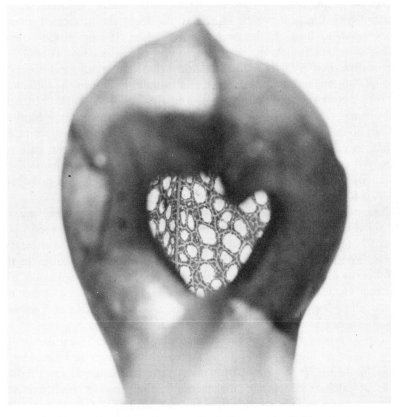

A close-up look into the mouth of the parrot pitcher plant.

ern states. But it does freeze out in winter and won't survive. At least, it hasn't yet in our attempts to acclimate it to New Jersey and Maine conditions.

When you look at this oddity close up, the aptness of its name becomes apparent. The hollow pitchers which are its leaves extend to a rounded hood with a bit of a crest down the center. At the tip of this hood, which curls back to face the center of the plant, is what passes for a beaklike point. Under it is a small mouth which is almost perfectly round.

Another feature that sets the parrot apart is the multitude of translucent spots on its hood. With adequate sun these spots become more pronounced and the reddish veins extend farther down the pitcher and into the wing portion of the leaf.

Lacking the spine-lined mouth of the purpureas, the parrot, *Sarracenia psittacina Michaux,* has evolved a somewhat different, yet equally effective, insect-confusing method of getting fed. The flared hood is up, the mouth toward the center of the plant. As insects are attracted into the tiny mouth by the aroma from the attractant glands, they begin to feed on the nectar of the plant. If they suspect something awry, the insects may attempt to flee. But which way is out? Not up, although there's light showing through those translucent spots. It may fly up, but merely hits its head and tumbles back into the pitcher.

How about the other three sides, left, right, and rear? The potential prey finds the same problem. Its apparent escape route is blocked. What look like openings are just those same translucent spots again. That leaves only one way to escape, providing the insect still has strength enough and the brain capacity to make the choice. Most likely many insects keep trying to fly up and out until they tire and drop back defeated.

The odds aren't much in the insect's favor. There's only one way out of six directions. Probably it becomes disoriented in the process and keeps banging its head against the same escape-proof inner wall of the hood. There's also the factor that more glands are placed lower in the pitcher to lure unwitting insects deeper into the narrower part of the pitcher. No wonder the parrot pitcher seems as well fed as its other relatives! You might say it stacks the odds aginst its victims and outwits them or just wears them down. Into the soup, that is.

The internal structure of this pitcher plant is basically similar to the others. Glands lower in the pitcher secrete the enzymes and fluids to digest the insects. It too grows from a central rhizome which can produce 4 to 15 pitchers depending how favorable the growing conditions are. This species is usually smaller than the purpureas. It fits better into terrariums with other smaller types: the sundews, butterworts, and flytraps.

Culture is almost identical to that of purpureas. It prefers slightly less water around it, so the middle part of the terrarium is preferable. When planting them as individual specimens or as a group of its own kind, follow the directions for purpureas. Just remember to give each a bit more drainage in the pot or tray in which you plan to grow it.

Allow me an aside here, please, since an ingenious growing method used by one carnivorous cultivator may save you time and work. After he has planted his single or several specimens in one pot, he waters them well and places the pot on a tray of gravel. There is always about an inch of water in the tray. Over the potted plants he places one of those wide-mouth glass gallon jars. (I hope he isn't seen too often entering and leaving the bars where he obtains the former containers of olives, cherries, and onions.) He assured me that in his work, traveling extensively week after week, this method provides insurance that the plants will have adequate humidity while he is away. Judging from the condition of the plants, it seems his system has some merit.

Parrot pitchers bear attractive red to crimson flowers on tall stalks, 8 to 15 inches high, each spring. Expect but one per plant, although larger rhizomes may produce two flower stalks. The flowers resemble those borne by purpureas, but about half the size. Although the coloration we've seen may be caused by brighter southern sun, parrot flowers have always seemed a deeper red, sometimes close to a reddish brown. Occasionally petals remain greenish red, while sepals become darker hued. Seed pods follow.

If you succeed in flowering several varieties of sarracenia at the same time, you may earn a bonus. They do occasionally cross-pollinate. You may discover the seeds will develop into a natural hybrid plant. For fun, you can attempt to hybridize these plants yourself by hand. Details are in the chapter on experiments.

Hooded Pitcher Plants

Hooded pitcher plants, *Sarracenia minor,* demonstrate a variation on the parrots' insect snaring ploys. They have an upright growth habit, but still emerge in the basic radiation pattern from the central rhizome. Most of the pitchers with their monklike hoods also face inward toward the center of the plant. Pitchers themselves may be 6 to 12 inches tall. During the growing season, you'll find pitchers of various sizes. These plants, like all others, continue to sprout new pitchers as older ones dry and die.

Hooded pitcher plants, *Sarracenia minor,* have mouths somewhat like the parrot pitcher, but have the translucent spots typical of the cobra lily and its upright growth.

The pitchers themselves are hollow, widening at the top into a rounded hood with the mouth on the lower inside portion of the hood. These plants also have translucent spots arranged across the entire hood and a short distance down to what might be considered the throat area, just below the plant's mouth. With strong sun, they'll change color, ranging from darker green to green with reddish tinges to reddish

This rear view of a hooded pitcher shows its distinctive translucent spots.

brown. Some hoods seem to have more of the translucent spots than others, but we haven't yet determined how or why.

When insects are enticed into the mouth, they feed on the nectar secreted by the inner cells of the hood, as they do with other pitchers. The attractant glands are arrayed around the rim of the mouth and inside as well to encourage the dinner prospects to venture farther. Once inside, the odds against escape are even worse than the parrot gives. The mouth, remember, usually is lower, facing the ground. We assume the insect's instinct tells it to fly up or to the sides, wherever the faint light indicates there may be an avenue of escape. Too bad for the insect, because there is no passage from its little plant cave. Eventually, it either falls into the digestive fluids in the lower portion of the pitcher or unwittingly crawls its way down to a watery grave.

Each spring, delightful yellow flowers, usually one per plant, arise on slender stalks. They are similar to the yellow-petaled flowers of the tall huntsman's horn, *Sarracenia flava*, but slightly smaller. Petals are somewhat shorter, too.

When you dissect the hooded pitcher, you'll find the same zones common to all pitcher plants. Botanists we have visited emphasize that there are normally five distinct zones, from the hood or mouth into which the attractant glands first entice the insects, down to the final base in which absorption of the digested protein and nutrients takes place. It may be difficult for the untrained amateur to recognize these distinct areas. In fact, they may really be sub zones, a different type of pavement on the path the insect takes from entering until it becomes the plant's nourishment.

However, in deference to the more technically minded, we have included with these chapters outline sketches that will help you identify the inner workings of the pitcher plants. These may be especially useful if you plan study projects and need a guide to re-create accurate drawings of these plants' innards.

Hooded pitchers will thrive with the same basic planting and care given to their relatives. In fact, the following friend of the same family, the miniature huntsman's horn, also is cultured in the same way.

Miniature Huntsman's Horn

This lovely little plant, the miniature huntsman's horn, goes by the scientific name of *Sarracenia rubra*. Why rubra, I don't know. Seldom does it attain the reddish hues which you might think it displays. With sufficient strong sun or ample artificial light it will become more reddish across its top and partially down its pitcher. In most instances, it remains overall green with fine reddish veins. We have subjected these plants to 16 hours each day of Vita-Lites and Gro-Lux, but have not achieved the darker red color we expected it to become.

This plant too emerges from a rhizome, which tends to elongate. Propagation is easier by breaking the rhizome in several spots and replanting the pieces. Pitchers in various stages of growth should mature 6 to 12 inches tall. It is identified by the slender pitchers, veined from halfway up with reddish veins more prominent and deeply colored as they reach the top. It doesn't really have a hood as such. The top of

A typical grouping of miniature huntsman's horns.

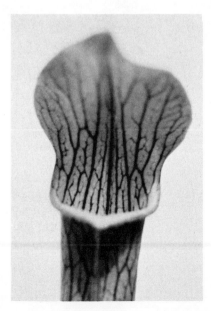

The mouth of the miniature huntsman's horn may be smaller than other pitcher relatives, but it's large enough to catch its share of insects.

the plant is distinguished by a slightly flared flap. Without translucent spots to confuse the insects, and with an open mouth that seems hardly sufficient to deter its prey's escape, you might wonder how this species manages to survive.

Judging from numerous dissections of these plants it does quite well at insect-catching. Perhaps the attractant is more powerful. Perhaps it is easier for insects to enter, especially those that may be suspicious of the other types. More likely, the glands that secrete the aromas to attract insects are stronger further down the inside of the pitcher. Of course, it does get narrower closer to the ground. That too may make it more difficult for insects once halfway down to turn around and head for home. If you cut one open, you'll find a multitude of insects lodged at various levels.

This cross section of a miniature huntsman's horn leaf shows its insect-catching ability.

Each spring this delicate pitcher plant bears single red, velvety flowers on individual stalks. Some wave several inches higher than the pitchers grow.

Miniature huntsman's horns should be grown as hooded pitchers are. They prefer sufficient moisture in the planting medium, but slightly less than purpureas and parrots.

The flowers of the miniature huntsman's horn appear
on tall single stalks. The blooms are smaller and lighter
in color than the deep red-purple pitcher flowers.

All of these pitcher plants enjoy such similar conditions that they are
easily grown together in terrariums. Remember, they do need room to
reach their full potential size. Select a large-enough display terrarium
for a group planting, with sufficient headroom so the taller types can
flower properly.

Each type has its own distinctive insect-alluring methods. Together
they make a fine display.

Huntsman's Horns, Sweet Trumpets, and Cobra Lilies

Among the tallest, biggest, and most striking of the easily grown and more readily available carnivorous plants are the huntsman's horns. Huntsman's horns are actually the tallest varieties of the pitcher plants. That includes *Sarracenia flava* and its many offshoots, cousins, hybrid relatives, and associated friends. Typical of this family are their tall, straight, openmouthed growing habits. Many of the family have flaps over the mouth to partially conceal the yawning hollow pitchers which are one-way streets for the unsuspecting insects who enter in search of the sweet nectar's aroma.

Huntsman's horn hybrids offer variations in color and shape, size and configuration. You have a wide selection for complete terrariums from the *Sarracenia flava* down to the miniature huntsman's horns, S. *rubra* (discussed in the previous chapter). When you begin counting the many variations in this carnivorous group it's like going to a family reunion. They all look alike, but it is difficult to recall all the names and places, never mind the true and accurate family relationships. Sarracenias are like that. Even the experts have trouble determining and defining the exact names and ancestral heritage.

Perhaps one of the most unique of the taller carnivorous plants is not even related to the Sarracenia family. Botanists still debate its true origin and exact name. Yet, the cobra lily, *Darlingtonia californica*, stands tall as one of the most interesting of the insect-luring plants. At least it does in North America, and ranks high among the more fascinating in the world as well.

The huntsman's horn, *Sarracenia flava*.
(Courtesy Burgess Seed & Plant Company)

I've put all the taller carnivorous plants in this particular chapter for a reason. For one thing, because they are tall they require extra space for growing to their destined dimensions; for another, because their cultural life styles are similar. And finally, because they have similar insect attracting, catching, and eating habits. After all, when you've studied several hundred thousand over the years, and observed so many more from coast to coast and overseas, you just naturally develop patterns with these carnivorous plant friends.

Common to all these taller carnivores is their trapping method. Insects are lured by nectar secreted by the plant glands near the mouth and throat. Some species have minute hairs lining the upper portion of the hollow stalks or trumpets. Others assist their would-be meals down the inside track by waxy surfaces. A few astound and confuse the intended victims by coloration or devious and perplexing surfaces.

Among the taller-growing carnivorous plants native chiefly to North America is the cobra lily. This plant is distinctly different and in a class by itself. Maturing some 18 or more inches high, it represents its name well. With flared hood, forked reddish tongue, and twisted stalk swaying in the breeze, this plant has gained new popularity in recent years. The cobra lily, native to the bogs of mountains in Northern California, Oregon, and other spots in America's Pacific Northwest justly deserves its fame. Anyone who has seen pictures of the king cobra snake will realize how this plant earned its name. Closer inspection reveals even more fascinating information.

The cobra lily, with its forked tongue, resembles the snake for which it is named. The plant's mouth is beneath the tongue, and the hood is composed of translucent spots which block an insect's attempts to fly straight up and out.

A group of huntman's horns growing in a marshy field.

All of the tall pitcher plants in the Sarracenia group grow from rhizomes, much like an iris does. All prefer the semi-swampy conditions which are found near bogs, stream borders, and coastal wetlands. All have hollow pitchers into which insect meals are lured to provide the nutrients required for satisfactory growth.

Some varieties have more colorful tops and hoods or lids. Others even have variegations and more colorful veining such as is apparent in the sweet trumpet, *S. drummondii.* In fact, some from careful observation seem to outdo their relatives in their ability to lure a higher proportion of insects than surrounding species. Specialists conjecture, or perhaps they just guess, that color is more appealing. Others believe the factor is a more powerful attractant in some species than in others, even though they may grow side by side.

Colorful reddish purple veins against a white background identify the sweet trumpet.

Let's see in detail how these plants go about their work. First, the insect is attracted by the nectar secreted by the aroma glands below the mouth of the huntsman's horn. As it flies nearer, it must obviously find the scent alluring. Why else would it trespass into dangerous grounds?

All sarracenias have similar mouths, wide open and leading to a hollow stem or stalk that becomes progressively narrow as it nears the base from which it grows. A flap or liplike hood is more pronounced on some species than on others, but occurs on all. That's a family resemblance you can spot immediately. Once the insect enters the open mouth it becomes aware of the additional attractants farther down inside the pitcher.

There are glands inside the area that can be considered the throat of the pitcher or horn which lures the insect deeper and deeper. At a point the insect may become alarmed. Some may escape. The majority, unthinking as they are, may wish to rest on the inner sides of the plant stalk. Unfortunately for them, the mid- to lower-inner surfaces are more

Diagram of a single huntsman's horn.

slippery. These lower waxy or greasy surfaces speed the final movement of the insect prey into the digestive fluids within the stem. Glands located in the lower-inner walls of the huntsman's horns secrete juices which combine with moisture brought up from the roots or deposited by rain to create a soupy broth within the pitcher. In this the unlucky insect becomes a meal for the plant.

Huntsman's horns attract and eat a wide range of insects each week. We've found flies, moths, crickets, grasshoppers, ants, and even small frogs in various stages of digestion within the pitchers we have cut apart.

At flowering time the huntsman's horns, cobra lilies, and their relatives put on a colorful show. *S. flava* and most closely related species, including the natural hybrids, send up tall stalks which bear the most curious, large, and showy yellow flowers. In the world list at the end of

Huntsman's horn flowers (above) are a bright yellow; the cobra lily's are a deep, dark red.

this book you'll find some examples of natural and man-made crosses which result in some color changes of the plant trumpets and flowers as well. In general, these plants bear flowers with large petals. The blooms may be two or more inches across.

Seeds are fairly large and offer good opportunity to propagate these plants from seeds. Follow the accepted sprouting method in mixtures of sphagnum and peat with sand as described in the culture chapter. Another simple way to reproduce the various huntsman's horns, *S. flava, leucophylla, oreophila, alata, rubra,* and *minor* plus their related hybrids, is by rhizome cutting. As the plants grow they build strength and growth in the underground rhizome which resembles that of your typical iris. These rhizomes may be an inch thick and quite woody, so cut carefully. Take a sharp knife and cut segments an inch or more long, leaving bits of root on the pieces and stalks or stalk buds. After a few weeks in moist sphagnum moss these active pieces should begin to sprout new plants.

The first leaves to appear are termed juvenile leaves. This trait is common to all the American species of pitcher plants. These leaves don't open, but do help get the plant growing and established. In fact, during fall, especially on young plants only a few years old, juvenile leaves may be found which last through the winter. They are also termed winter leaves. In spring the more desired hollow leaves or pitchers grow. The real activity, that fascinating insect-catching ability, is accomplished by these mature leaves, which are hollow. Some may be only a few inches tall, on some species like the miniature huntsman's horn, *S. rubra.*

As the pitchers of each variety mature they may in taller varieties reach 2 feet high. The mouths on these plants are covered or partially covered by flaps or hoodlike lids. Some mouths may be less than ¼ inch in diameter, up to perhaps ½ inch wide. Others, in *S. flava* and the sweet trumpet, *S. drummondii,* may be 1 to 3 inches across.

Despite these differences in size, shape, and coloration, all pitcher plants have the same five zones inside their hollow pitchers. Since these taller types have fewer hairs or spines lining their mouths and throats, perhaps we should review the zones. They are detailed in the earlier chapter as they apply to the purple pitcher, hooded, and parrot species.

The first zone includes the opening plus the underside of the flap or hood as well as the area just under the lip or rim of the pitcher. This is the attraction zone. It contains the glands and, depending on variety, stiff hairs. The glands secrete the aromas which lure the insects from fields away to their imminent demise. The hairs, on those which bear them, also contribute to directing the insects into and down into the lower areas of the pitchers.

Huntsman's horns can reach from 18 inches to three feet tall.

Botanists list zone two as similar, with fewer or no hairs, and gradually changing into zone three. This zone is characteristically lined with a waxy surface. If an insect tries to gain a foothold here he is out of luck. Other carnivorous specialists refer to this as the sliding or slippery zone.

Below this are usually the digestive glands, which secrete the enzymes to break down the insects' bodies to be absorbed by absorption glands in the lower zones. The liquid inside the pitchers, brought up by water-conducting tissues and also deposited by rain, usually remains level to about zone three. You can examine the inside of a pitcher merely by slitting it. That's another project for younger hobbyists. Just

An insect's view into the deep throat of the huntsman's horn.

cut the entire pitcher open, use pushpins to hold it open and then identify the different zones and their functions.

Actually, different species do vary in the size and complexity of their
catching and digestive abilities. But basically, the zones can be seen, especially by use of a hand magnifying glass.

Over the years we have learned that all these huntsman's horns do
thrive in sphagnum moss alone. However, since their original habitat is
a somewhat drier, acid, and sandy soil, you can utilize mixtures of these
materials in the bases of large growing terrariums and containers.

Sarracenia flava, the most common yellowish-green huntsman's horn,
is attractively ribbed toward the upper areas and the mouth. As you
begin to classify and collect the various species and hybrids you'll find
that colors do change. On some species reddish hues become apparent
and more pronounced, even blotches at the rear of the throat.

For strikingly lovely coloration, the sweet trumpet, S. *drummondii*, is
perhaps the most attractive of the huntsman's horns. Tallest of the
huntsman's horn family, sweet trumpets are distinguished by variegated reddish to purplish necks, throats, and mouths, including their
flap over the mouths. They also have unusual translucent whitish areas
between the ribs and veins which make them especially appealing in a
combination terrarium planting. This is one type which bears red
blooms, rather than the more typical yellow or yellowish green. Al-

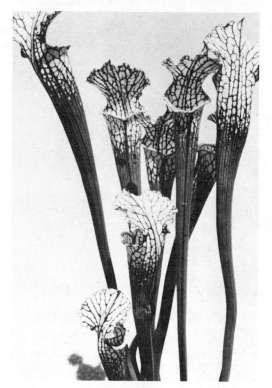

A potted group of sweet trumpets adds a colorful touch to a collection of huntsman's horns.

though for years this sweet trumpet has been living under the Latin
name *S. drummondii*, botanists now wish to change it to *S. leucophylla*.
(Somehow, I like the sound of *drummondii* better.)

Among hybrids, the crosses between sweet trumpet and others often
yield the most striking coloration, both in foliage and flower. Perhaps
one of the most dramatic results came from our crossing sweet trumpet
with the common northern pitcher plant, *S. purpurea*. The offspring
have a more upright growing habit, the exceptional coloration of the
sweet trumpet parent, and they respond well to terrarium and individual cultivation.

The natural range of huntsman's horns is the southeastern part of
the United States.

Sarracenia flava ranges from Virginia through the Carolinas into
Georgia, Alabama, and Florida, in bogs, swamps, and moist areas of
semi-shaded woods and pinelands. The tall summer leaves may grow be-

This hybrid has the coloration, upright growth, and
top flap of the sweet trumpet and the open mouth
and spiny throat of the purple pitcher.

tween 18 and 30 inches tall. Flowers are bright yellow in good sun to
yellowish-green in shaded glades.

S. *alata,* formerly known as *sledgei,* is more restricted in its range to
the southerly areas of Alabama and Mississippi, even into parts of
Texas, where bogs and low moist areas favor its growth. It is lower-
growing than flava and has more red coloration in the upper portions of
the pitchers. Flowers borne on single stalks are typically yellow-green.

S. *oreophila* occurs in Alabama and Georgia, with some in Florida,
too. It seems to prefer sandier soil but does well in acid bog soils which
were probably once peat moss. The leaves rise about 18 inches high, are
green, and the flowers typically yellow-green.

Sarracenia rubra is the lowest-growing type of the upright pitchers. It ranges through the Carolinas to Florida and west to Louisiana in bogs, swamps, and roadside ditches. Although most pitcher plants have from four to a dozen pitchers per rhizome, this smaller variety may have several dozen pitchers in various stages of maturity from one rhizome. It is more usually a clump-growing plant, reaching 6 to 12 inches tall.

The top portions of rubra leaves are, as the name indicates, reddish to dark red in the sun, and its flowers are deeply red or maroon, smaller but similar to the purple pitcher plant blooms.

Cobra Lily

Just as tall, but with much more eye appeal to people and to insects as well, I suspect, the cobra lily can rightly be called a pitcher plant.

After careful study, botanist J. D. Drackenridge, who is credited with its discovery, decided there were some striking differences between it and other pitcher plants, so he christened this outstanding carnivore *Darlingtonia californica*. However, after proudly bearing that name for many years, it has been declared by modern botanists to be invalid according to the rules of botanical descriptions. Thus, *Darlingtonia californica* has become *Chrysamphora californica*. Perhaps the common name of cobra lily or cobra plant is more descriptive and appropriate.

Looking at this strange plant you can easily understand why the common name was so readily applied and accepted. The plant arises from a rhizome with juvenile leaves much as pitcher plants do. True, it also has the similar attracting and digesting zones inside its hollow pitchers. There the similarity ends.

The top of each mature pitcher is flared into a hood which carries over and around the front. Attached to this front section is a forked appendage, much like the forked tongue so typical of snakes. Beneath this tongue is the somewhat hidden mouth, oval-shaped but just as hungry as the open yawning mouths of other pitcher plants.

The hood and part of the pitcher itself is covered with translucent spots between the veins. The tongue itself is greenish red to dark maroon.

As an insect is attracted to the plant it may light first on the tongue or fly directly upward into the open mouth. The interior attractant glands secrete their nectar to lure the victim lower where it can become a meal. But sometimes insects have a change of heart. Flying upward is no answer inside this carnivore. The hood covers the entire escape route on almost every side. Besides, flying up to the light just means a

 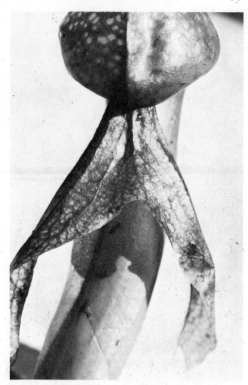

Diagram of a single cobra lily.   The reddish forked tongue of the cobra lily lures insects, and also serves as a walkway into the plant's mouth.

bumped head for the insect. Every spot that seems an escape door or window is securely closed. Eventually the insect drops exhausted into the liquid to be digested in its turn.

Each summer the cobra plants send up several flower stalks. These look similar to the Indian pipes you find on woodland walks, but are greenish with some reddish coloration near the top. The blooms are an attractive dark red, between the size of the large huntsman's horn flowers and those of the miniature huntsman's horns, about 1 to 2 inches across and velvety in texture.

Propagation can be accomplished from seed in a sphagnum moss mixture sprouting tray or container. It is easier to take root or rhizome cuttings, although with care you can often succeed with cuttings of young, strong pitchers carefully kept moist in sphagnum moss with just their tops sticking out.

Repotted from a parent plant, this cobra lily seems to stand watch over a new brood of newly sprouted youngsters.

Cobra lilies are native originally to the mountains of northern California and Oregon along the coast. They are conspicuously bog-loving and require much greater humidity for successful cultivation in terrariums or other containers. In fact, although they will grow as potted specimens, it pays to surround them with glass or plastic hoods to provide that essential humidity these plants crave.

This 9-inch-tall clear terrarium lets you grow the exotic cobra lily in a humid atmosphere and view the full range of its growth, from juvenile plants to mature hoods and flowers.

Huntsman's horns in all their myriad sizes, shapes, and colors—and their most-likely distant cousin the cobra lily—offer unique specimens for larger terrariums. Their height requires lots of growing room, but grouped together in your collection, they'll provide hours of pleasure as they exhibit their own individual life styles and insect-catching abilities.

Butterworts and Bladderworts

Butterworts and bladderworts have the same hungry insect-eating habits as other carnivorous plants.

Bladderworts, although mostly quite small, are primarily aquatic. True, a few are terrestrial, but most live their catching and dining lives in water. They appear quite active with their ability to seemingly vacuum tiny aquatic insects right into their digestive bladders, but since they are so small, and less easily grown, we'll leave them to the last half of this chapter.

Butterworts rate perhaps greater attention because they are quite easily cultivated. They fit into terrariums, bear lovely flowers, and provide yet another insect-catching method for observation. Being larger, they also can provide extra color and display in a carnivorous collection.

Butterworts are members of the Pinguicula family. You can find many roadside ditches and swamps across America in which butterworts lurk, patiently watchful for small insects to alight on their sticky leaves. They are primarily native to the Americas and range in size from a dime to some species that may measure 6 inches across. There are some 35 species distributed across the Northern Hemisphere, even into Scandinavia. From the tiniest *Pinguicula vulgaris*, with leaves smaller than a child's thumbnail, to the larger types, P. *lutea* and P. *caudata*, you can grow an array of butterworts quite easily. The largest may easily mature up to 5 or 6 inches in diameter.

These odd little plants aren't really very active. But they make up for that fact in the strongly adhesive properties of their leaves. Like so many plants, butterworts are well named, when you consider their true Latin name, Pinguicula. It means somewhat fat. The leaves have a waxy or greasy appearance and appear harmless enough. But the secre-

Butterworts in their native habitat. (Courtesy Carolina Biological Supply Company)

tions of the leaves on the surface and along the tiny hairs of the leaves have surprising strength. When unwary insects are attracted to these oval or tongue-shaped leaves they may suspect no danger. But once they place their tiny insect foot on the sticky leaves they may realize the dilemma. Trying to flee, the insect places its other feet on the sticky surface. Poor pest. It learns soon enough that that lovely resting spot has epoxy gluelike ability. The more the insect struggles, the more surface it contacts.

Larger plants are more obvious in the manner in which the leaves actually roll to clinch the decision and capture its meal. Perhaps the leaf really does slowly roll toward its potential meal. Some experts believe the leaves are stimulated by insect activity into a faster growth pattern that results in the leaf furling. More likely the insect, in its struggles to escape, contacts more leaf surface and the sticky substance helps roll the leaves around the victim.

This large butterwort (the quarter shows its relative size) can fill a planter by itself.

The sticky, hair-lined leaves of the butterwort. Notice how the leaves furl when insects are caught.

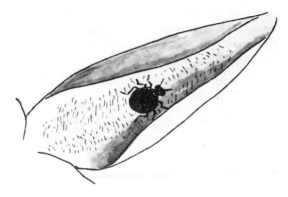

The leaf-furling habit.

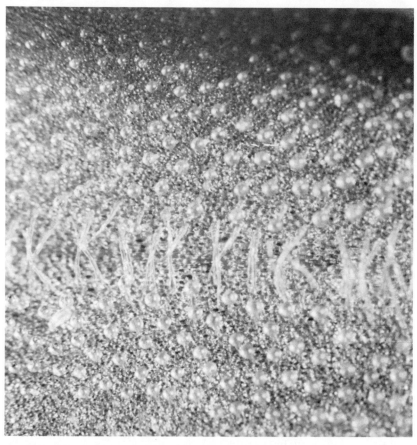

Close-up look at the mucilage glands of the butterwort leaf.

Close observation reveals that the leaf has incurled margins. This forms a shallow catching pan or surface which is covered with colorless secretions. There are two types of glands distributed on the leaf surface. The first is visible and appears stalked or hairlike. The second type is composed of groups of smaller cells which, under a microscope, appear as knobs on the leaf surface. This type secretes the digestive enzymes. The first type is responsible for the sticky catching substance.

Although they appear passive, butterworts do demonstrate some primitive nervous system understanding. Their ability to distinguish between rain, artificial stimulation and the real thing—insect meals—becomes obvious after you watch them in action. For example, when you water these plants the leaves just don't curl inward. Even when you stimulate the surface with a pencil point or other object, the leaves remain basically in the same position. But let an insect land and the plant gets the message that it's mealtime. Additional sticky material is secreted to secure the insect tightly. As this process continues, the insect's struggle stimulates release of digestive enzyme from that second, less-conspicuous series of glands. Within about a week the nutrients are absorbed from the insect to nourish the plant.

Even the smallest and medium-size butterworts catch a surprising range of insects. Although we can't actually attribute special taste preferences to these plants, gnats and fruit flies seem to be the most common insects in their diets. In fact, growing cultures of Drosophila (fruit flies) in a terrarium with butterworts will keep them happy for months to come. Fruit flies also are easily caught and utilized for food by the many smaller sundews.

Although the curling of butterwort leaves is obvious after a day or so, younger plants can accomplish the process in several hours. One aspiring filmmaker set his Super 8 camera on stop action so it would take a frame every 16 seconds. After some testing, he achieved an amazing result. When the film was played back at regular speed he had photographically captured the action just as effectively as the butterwort leaf had captured the insect.

Flowering is another attractive bonus from butterworts. The smaller so-called swamp violet types bear single tiny violet and lavender flowers on short stalks. The larger varieties bear blooms of violet and in some species yellow which may be almost an inch or so long. The form is much like the common wild violet found in woods and fields. There are five sepals, two stamens, and a compound pistil on each flower.

There's another fascinating sidelight to the butterworts. Some, particularly the *Pinguicula pumila*, *primuliflora*, and *planifolia*, may develop baby plantlets right on the margins of their leaves, all by themselves. Actually, almost all butterworts can be induced to reproduce vegeta-

The butterwort blossom.

tively this way by cutting off the leaves and placing them in a propagating bed of sphagnum moss which is kept moist at all times. Just bury the lower, cut portion of the leaf in the moss and within several weeks new baby plants should begin to form.

You also can reproduce butterworts if you save seed and sow it over damp sphagnum and peat moss, covered with plastic sheets. Germination may be rapid or require several weeks of real patience until they sprout. The northern species, especially *P. vulgaris* and *P. villosa*, usually respond better if seed is stratified or put into near-freezing conditions for several weeks. The southern species, since they seldom face natural freezing in their usual habitat, sprout without this extra step.

To culture butterworts, follow the methods for sundews. Soil, however, should be allowed to become a bit drier between waterings. You can satisfy that requirement by growing butterworts separately or in a slightly higher portion of the terrarium.

Another point: butterworts do also require a period of rest, although they are not a bulb or rhizome type plant. You can allow them to dry down a bit for a month to 6 weeks after they have flowered and set seed. After the resting period, during which they should be only sparsely watered, you can bring them back into full growth by applying the usual waterings several times per week.

Under lights, the plants require semi-sun conditions and flower and set seed best under double Gro-Lux or Dura-Test fluorescent tubes 18 inches above the plants.

The list of butterworts found in the worldwide carnivorous plant list provides locations for the different species. However, it seems proper to list some descriptive information on the more commonly available types here for reference in your trips afield and for use in growing these easier ones successfully.

Pinguicula villosa ranges across Canada and into Alaska in sphagnum and peat moss bogs. It bears single lavender or bluish flowers per plant. The leaves tend to be circular, rather than elongated or tongue-shaped, about ½ to 1 inch long.

Pinguicula pumila is found in the Carolinas and southeastern states, including Florida, in open pineland and sandy moist soils. The flowers range from white to violet, some with yellowish tongues. The leaves are oval and light green, from ½ to 2 inches long.

Pinguicula pumila Buswellii is similar except that it has yellow flowers.

A common northern variety, *Pinguicula vulgaris*, inhabits bogs and light soils from New York and the New England states across the northern tier of Michigan to Montana, northern California, Washington, and Oregon, as well as in Canada. This butterwort has one or more violet flowers, slightly larger than tiny wood violet flower size. Leaves are tongue-shaped, even to elliptical shape and normally light green. They may be an inch to 2 inches or slightly longer.

Pinguicula lutea is a butterwort found in the Carolinas and southern states. It enjoys deep sandy soils and bogs as well as open pine woods and is perhaps the most common of the butterworts. Flowers are bright yellow on stalks 4 to 8 inches tall which can be seen above roadside grasses. Leaves are distinctively yellow green and rounded, like a miniature football that has been deflated with curled edges. The leaves can be 1 to 3 inches long on mature plants. We've grown some plants 6 inches in diameter.

Pinguicula caerulea, planifolia, primuliflora, and *ionantha* are other types found mainly in the southeastern states. The ionantha is mainly native to Florida.

Although all butterworts have the same basic growth and culture re-

The southern butterwort, *Pinguicula caerulea,* is native to the great Green Swamp of North Carolina but is also found in other parts of the South. (Courtesy Longwood Gardens)

quirements, they do resemble each other to the untrained eye. Flower color and size are more exact means of identification, but then that's the province of more detailed botanical textbooks.

Going further afield for the fancier who aims for a captivating collection, the *Pinguicula caudata* from Mexico and Central America is exciting. It has large yellowish to reddish leaves with showy, purple flowers.

The subspecies *P. mexicana* is native to Mexico and may grow up to 4 inches or more in diameter. Most likely a variation on the *caudata,* the leaves are usually almost round with a slight lip along the edges.

However, during various times of its growth the leaves may take on different shapes. Size is the best identifying characteristic of these two types. Mature plants can grow up to 8 inches across.

Bladderworts

Among carnivorous plants, perhaps the bladderworts, scientifically the Utricularia family, have more types and styles than any of the others. More than 200 species have been identified and named by bota-

This bladderwort has been lifted from the water to a plain white surface to show its typical rosette growth pattern.

nists and specialists in all parts of the world. There are about 20 found native to the United States.

Although these are mainly aquatic, their habits, especially among the larger types, bear watching. For one thing, they have been called fish-eating plants. Bladderworts look harmless enough, but can they do tricks. In nature they are found in bogs, swamps, along still lakes and waterways. You might never realize their insect-eating abilities unless you took a close underwater look at them.

Bladderworts have characteristic threadlike growth and may be attached to either the mud or the soil at the bottom of the pond or brook. Some are just free-floating. We've explored old cranberry bogs so covered with these plants in spring and fall that there seems a slight pink or lavender haze over the water from the thousands of tiny blooms.

There are several types which thrive at the edge of ponds and bogs. After rains they may be submerged for a time, but then they either break loose to float freely or grow to the surface on a slender, threadlike stalk. The flowers have two segments, usually called lips. Like the butterworts, these plants have a palate between the two lips. Stamens and pistil are sheltered inside the petals. Although the flowers are delicate, without them the different species resemble just so much threadlike algae floating on the water.

But there the difference ends. Beneath that surface is the action. And what action! Although different species may be larger or smaller, with bladders somewhat different in shape and position on the plants, they share one common ability. Perhaps it should be called an uncommon ability, since these plants do lure, catch, and eat insects. In fact, these can rightly be called carnivorous. In addition to insects, they will eat small crustaceans, worms, fish, diatoms, and protozoans.

Basically, all bladders are somewhat oval or egg-shaped. Early studies reached the conclusion that the bladders were for flotation of the plants on the water. That was accepted until it was discovered that the bladders provide a more insidious function. They are highly active traps not filled with air for flotation, but with a vacuum. As tiny water insects or animals touch certain parts of the bladders near the so-called mouth, the trap or lid opens inward. Water rushes in to fill the vacuum. In the process the insect, water flea, or baby fish is swept right along. Once the space occupied by the vacuum is filled, the trap or lid closes and digestion of the imprisoned meal begins.

We've taken time to closely observe this mysteriously effective insect-trapping mechanism. Once you understand how it works, you too can watch in fascination as these plants perform their wonders for you. They are easily grown in gallon jars or fish tanks, so observation through the clear glass sides is easy.

Naturally you'll need good eyesight since these trapping bladders are sometimes nearly microscopic, the largest about half the size of a pencil eraser. Persevere and you'll be well rewarded.

The entrance to the trap or bladder is circular, on one side of the bladder. The closing flap, or trapdoor if you prefer, is a sheet of plant

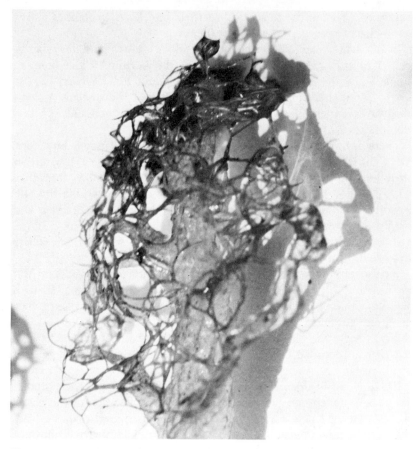

Flotation isn't the only function of the bladderwort bladders. They also lure and eat insects.

tissue attached to the top of the bladder, much like a spring hinge, opening into the bladder. This trapdoor hangs down from the top and is held in place, scientists believe, by cell pressure and parts of the plant. The surrounding portion of the bladder acts much like a door-stop which helps seal the bladder until the vacuum is built and the trap is cocked and ready for action.

Like the flytrap, bladderworts have trigger hairs. Unlike the flytrap, these are on the outside of the trap, usually just below the entrance of the door. They also help keep excess algae and material from entering the interior of the bladder when it is triggered into activity.

Nectar glands are also located just outside the door; right at the threshold, so to speak. As nature sets the stage, the trap or bladder sides

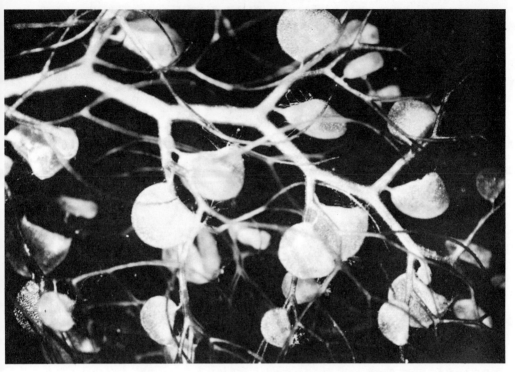

A closer look at the bladders. (Courtesy Carolina Biological Supply Company)

become concave. That's when the fun begins. As tiny water fleas and other microscopic potential meals approach, attracted to the secretions of the nectar glands, they touch the triggering hairs outside the door. With a quick response these hairs telegraph the approach of an imminent meal to the bladder. The doorstop gets the message and moves sufficiently for the trapdoor to be released. It swings in and up.

The water carrying the unsuspecting insects or tiny animal life rushes in. As it does, the trap itself fills, assumes a convex shape, and the door is pushed closed. Other tiny hairs or stalks inside the mouth of the trap help prevent the insect's escape. Since the water pressure is now equalized and the struggles of the victim further stimulate the bladder to secrete digestive juices, the door remains shut until the meal is done. Then, in nature's marvelous manner, special glands inside the trap absorb the water inside the bladder. This causes the vacuum which again sets the door securely in place and prepares the bladder for its next luring, catching, eating sequence.

An even closer look of the bladderwort leaf trap. (Courtesy Carolina Biological Supply Company)

It would seem to many that if there is a vacuum created inside the bladder the water would push the door open. Actually, close scrutiny has revealed that as the vacuum is formed when water inside is absorbed by the glands, the convex shape imparts an engineeringly sound structure to hold that door in place. Only stimulation of the outer trigger hairs can and does unlock the trapdoor again.

When no meal is caught, the digestive process may take a day or several. If no meal is swept inside, the inner glands simply go about their business in cooperation with the intricate mechanism of the plant to reset the trap. That can be done as soon as 15 to 20 minutes after the artificial triggering of the trap.

In spring, when tiny larvae, mosquito wigglers, and newly hatched fish are present, the traps earn their keep. Even somewhat larger meals may be sucked inside with just a piece of tail left outside the door. Propping open of the door may cause the entire bladder to die, or, at least, not perform its function properly.

Don't worry about your plants. Each has many other bladders to take over if a few pass on. After a bladder has been successful several times, the remains of its meals become even more visible. Eventually these skeletal parts fill up the bladder and it drops away from the plant stalk. Others, naturally, grow to continue the process. Any individual plant will have many, sometimes dozens, even up to hundreds of bladders of various sizes. It's just nature's way of replenishing the parts which make up the needed mechanism for these inspired aquatic insect-eating plants.

If you enjoy photography as well as growing plants, you can add to the pleasure of both hobbies with a camera and telephoto lens. Actually, a macro lens or bellows attachment will let you focus quite closely, about 4 inches or so away from the scene of action. But special lenses permit you to magnify the tiny bladders to capture an insect's-eye view of that moment of truth when the trap is snapped and the plant's meals pop inside.

Since these are small carnivorous plants, a microscope comes in handy. By bottom illumination and a bit of dye from school science kits, you can follow the action just like on closed circuit TV.

Bladderwort varieties are available from several firms. Most, however, don't ship except in spring and fall. If you want to find your own plants, you can usually spot areas of these amazing wonders in ponds or water impoundments in spring and summer when their thousands of flowers mark their presence. To collect them, take along some plastic freezer containers or large, sturdy plastic bags. A fine unit is a gallon plastic mayonnaise jar available from restaurants.

To grow these oddities successfully, we found it pays to put an inch of fine gravel on the bottom of the tank. Then add another layer of sand. Pond mud works, but too often muddies the water during movement of the container during study. Float the plants in several inches of water, allowing room for them to send up flowering stalks. These can be several to 6 or 8 inches tall in season.

Next, give them sunny areas or good lighting from artificial sources so that the plants can begin to form new bladders. Don't worry if you clip or break some of the plants in transplanting or setting up your growing tank. New parts and, in fact, new plants can and often will form quite rapidly even from those broken parts by vegetative reproduction.

Suppliers normally ship bladderworts in plastic containers so they don't puncture in the mail. When they grow and flower you can collect seeds to sprout in wet sphagnum moss. As they grow, just remove the moss and place the new plants in the desired tank.

Bladderworts also reproduce asexually or vegetatively by several methods. One is by winter buds called turions. When you obtain or collect these and store at 40° to 45° F. these buds remain tight. When floating in water with sufficient sunlight and warmth, these buds will unfurl to grow stalks, bladders, and complete plants. Some produce what amounts to lateral buds between branches which can be cut with a bit of stalk to sprout whole new plants.

When you plan to search for these tiny carnivores afield, here are some of the more common ones and where they are likely to be found. Also consult the world plant list for locations of these and other carnivorous plants.

Utricularia purpurea ranges across most of the United States. You can find it in lakes, slow streams, and roadside ditches. It is distinguished by several purple flowers on a single stem. In large groups the area seems to have a purplish haze above the water during blooming season. This particular species has whorled branches circling the main stem, much like spokes on a wheel.

The *Utricularia inflata* is native to coastal states between New Jersey south to Florida in ponds, lakes, and slow water. It has several to a dozen yellow flowers on one stem.

Utricularia olivacea is found in southern Florida in shallow ponds and swamps. It seems to have only one or two yellow flowers per stalk with alternate branching. This one is so tiny, it takes careful observation to find it.

In northern areas, *Utricularia vulgaris* can be seen in the northeastern states across the country to Washington and Oregon. It also likes ponds and roadside ditches. You'll identify it easily from the 10 to 20 yellow flowers on the stalks which rise from several inches to 8 inches high. Look for tiny bristles on leaf margins to complete your identification.

Utricularia intermedia with its thin, flat, and somewhat serrated leaves is found in ponds and ditches in the same area as the vulgaris. It has several yellow flowers with spurs on the blooms almost as long as the tiny flowers themselves.

Utricularia minor also inhabits still waters and bogs in the northern tier of states. Similar to intermedia, it has several yellow flowers and leaves that are thin, flat, and smooth except at the tip end. This type has leaves and bladders on the same stem and branches. Intermedia seems to have developed a division of labor. Some stems have leaves, others the bladders. That's a key distinguishing feature of this species.

With 200-plus species and subspecies, it is impossible in the space available to attempt describing them all. These are, from our experience, the more readily found or purchased from the carnivorous plant firms who specialize in the weird botanical genus.

This bladderwort, *Ultricularia vulgaris,* has a fernlike growth.

If you prefer to stay with terrariums and forego the aquatic types of carnivorous plants, two terrestrial species are suitable for the soggier part of the container.

Utricularia cornuta is common to the eastern United States in peat and sphagnum moss bogs, even in wet shorelines of ponds and streams. It bears several small yellow flowers on stalks a few inches tall.

Utricularia resupinata ranges throughout the Atlantic and Gulf coast states. It is also found, for some reason, around the Great Lakes. It prefers shores of ponds and ditches. Unfortunately, it only offers a single or at best a few small purple flowers.

Utricularia subulata is common along streams, ponds, and boggy areas of the Atlantic coast and in the southeastern states. It has small yellow flowers on each stalk.

Butterworts and bladderworts are two more oddities that can add new carnivorous dimensions to your collection. You certainly have a wide-enough choice among these two types.

Nepenthes

Nepenthes, those insect-eating pitcher plants of the Old World, are in a class by themselves. They deserve to be. Early explorers of the islands of the Indian Ocean, Asia, and in jungles of those faraway countries with strange-sounding names were undoubtedly responsible for some of the first stories of man-eating plants.

Most likely they had seen some of the wild and weird nepenthes in action. That shock might lead anyone's imagination astray. In old botany books and documents, the first references to pitcher plants involves the nepenthes. During the centuries since, naturalists and botanists have classified approximately sixty species, all related to what they call the family Nepenthaceae.

Today the art of classifying plants has reached a more advanced stage. Many subspecies of the Nepenthes family have been described. In all likelihood, some of these are simply natural crosses or hybrids from species which grow near each other and have cross-pollinated. Perhaps some have just evolved differently.

In any case, there is no mistaking these curiosities. Their typical growth habit certainly marks them as strange specimens indeed. And like many things from the Old World, especially from the remoter parts of Asia, these plants retain their own mystique.

Some nepenthes appear to have a pitcherlike form. The majority bear pitcherlike structures at the ends of regular leaves. The leaves themselves appear to be much like the typical lush, broad leaves of other tropical foliage plants. The pitcher, of course, makes the difference. That's where the plants perform their carnivorous feats.

With the upsurge of interest in plants over the past decade and focus by many on the carnivorous types, nepenthes are becoming deservedly more popular. They are, however, still difficult to obtain for most hobby collectors, although arboretums and dedicated specialists are having success in obtaining different varieties for culture and display.

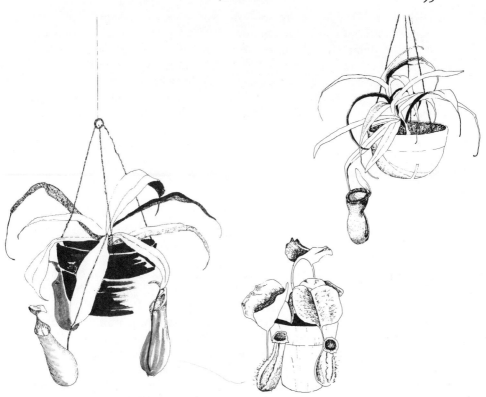

Some varieties of the Asian pitcher plant, nepenthes.

We've included some illustrations of the more easily available and readily grown varieties.

Among the sixty odd species native to tropical Asia, many are found in the Philippines, Malagasy Republic, and Indonesia. There are varieties in India, across parts of Australia and on other islands in between. Borneo and some of the other more exotic-sounding lands of Asia abound with the astounding carnivores.

Nepenthes are somewhat shrubby plants. They creep or climb on surrounding vegetation. You can, of course, grow them individually in pots or terrariums, providing you can supply the special conditions which match as closely as possible their preferred environment. Most nepenthes are jungle plants. That doesn't necessarily mean they need bright sun and torrid heat. After all, they thrive in the lower stories of the rain forests and jungles, so filtered light or semi-sun will meet their needs. Most species must, however, be provided with extra high humidity. When you examine them, that need becomes readily apparent.

High humidity and lots of moisture are required to supply sufficient liquid to the pitchers through the tendrils which attach them to the leaves. Without that liquid, the plants just can't produce the attractant fluids and digestive juices necessary to catch and eat their meals.

There are some species in this remarkable family of carnivores which do require less moisture. That's because they normally inhabit the somewhat drier hillsides, mountain slopes, or sandier bogs in their native countries.

Climbing nepenthes, compared to the ground creeping types, perform that function with tendrils. These are elongations of the leaf, which help the plant grow out and grasp supports so they can cling to the more desired locations for their best growth.

Although there are many variations in the pitchers, in size, shape, coloration, with modifications of mouth as well, all nepenthes bear a strong family resemblance. The pitchers are developed at the end of the tendril. Some tendrils may be short, others a foot or more in length. The pitchers also vary in size, from an inch or so to over a foot long. On individual plants you can find several sizes of pitchers in various stages of growth and maturity. In this respect they are much like the New World pitcher plants that sprout pitchers in continuing growth cycles to replace older ones which mature, dry, and die.

Between species there are striking differences in shape and size. Some may be long and narrow, with a wider top or bottom. Several are more like long, hollow cylinders from a few inches to almost a foot long. They may have caps, hoods, or fringelike mouths. Others include spinelike hairs around the entrance to the pitchers. Some of the largest pitchers may grow to nearly 18 inches long, about 4 inches in diameter.

These larger varieties are the ones that most likely led to stories of animal-consuming plants in those mysterious, faraway jungles. They can, in fact, catch small birds, rodents, and animals, as well as the more typical diet of insects. Remember, insects in the tropics are often quite unusual. Many are much larger than their distant relatives living in more temperate climates around the world. In consequence, I suspect, the nepenthes have developed their larger pitchers in which to catch their larger meals. Although there have been verified accounts of rodents, birds, and small animals caught in the larger nepenthes pitchers, most seasoned observers and professional botanists believe these plants intend to attract and devour only insects as their staple diet. Birds or small animals that have been found inside large pitchers more than likely entered to go exploring and drowned, or were affected by the anesthetic effect of the plant secretions, or just didn't have the room to flap their wings and fly out again from the restricted space inside the pitcher.

Most nepenthes trapping devices are formed at the end of long stemlike portions of the leaves. Here are three examples (left to right): *Nepenthes x superba* with an 8-inch pitcher, *N. maxima* with an 8½-inch pitcher, and *N. x dickinsoniana* with a 10-inch pitcher. (Courtesy Longwood Gardens)

Despite the obvious and often strikingly different appearances among the species, closer examination reveals they all have much in common. That fact is emphasized in their insect luring and digesting apparatus. First comes the mouth. That's logical, since the plant must have an opening through which unsuspecting insects are lured inside the pitchers. This may be a relatively small opening leading to a more bulbous "stomach" area or base. Or, it may be surprisingly wide.

Nepenthes rafflesiana pitcher. Note the "teeth."
(Courtesy Longwood Gardens)

Generally the mouth has a lip or rim. It usually has a flap or hood, which varies in size by species, of course. The rim or lip is composed of the turned-out margin of the pitcher, usually with some particularly distinctive corrugated, spiny, or fringed edging. In some species the ridges or fringes appear much like teeth. Again, these odd-looking plants certainly could stimulate the imaginations of those who first saw them, leading to fables and folklore stories of monster plants. Gin and tonic was a favorite formula in the tropics, and seeing these plants after adequate amounts of gin and tonic could easily expand imaginations, especially when recounting tales back in jolly old England at the explorer's favorite club or pub. After all, we all can improve a story each time we tell it to a new group, right?

Over the mouth, the flap, hood, or lid may range from tiny to so large it overshadows the mouth. For your reference this appendage is

called the operculum. The mouth, technically, is known as the peri-
stome.

Either just below, in, or on the lip (depending on the species) lies
the nectar gland. Some seem to have a large single gland. Other botanists report finding several.

The lip, whether fringed or smooth, offers no useful footholds for insects attracted by the secretions of the nectar glands. Some topple over
the edge into the fluid contained within the pitcher. Others are lured
farther inside to taste the sweet nectar on the inner area of the rim.
Tough luck. There's seldom foothold inside. The last sound the insect
makes is "splash," as it drops into the soup below.

For insects with strong grips, these nepenthes have another method
to capture their prey. Just below the rim and nectar glands is a smooth,
waxy surface, called glaucous by botanists. Consider it a slide for life,
right to the digestive lower region of the pitcher. This so-called con-
ducting surface is the second basic trapping method of nepenthes.

Below this region is the area of detention. In some species there may
be tiny hairs which thwart the insect's escape. In others, overlapping
epidermal cells are found. This lower area is where digestion takes
place. As insects fall, slide, or fly into contact with these lower, nether
regions, the digestive cells which line the walls of the pitcher begin to
secrete more fluid. The greater the stimulation, the more they produce
the juices which begin the breakdown of the soft nitrogenous parts of
the insect bodies. On close examination, scientists learned years ago
that these digestive glands produce a weak acid as well as digestive or
proteolytic enzymes. Bit by bit the process continues. The products of
digestions are absorbed into the interior walls of the pitcher, nourishing
the plants.

These basic regions inside the various nepenthes species are generally
similar. There are variations, of course, as there are in all related plants.
But, whether you cut apart one of the almost thimble-sized pitchers of
the smallest type or dissect the giant variety, their basic luring,
catching, eating methods are alike. What's more, they do work rather
efficiently.

There are some small insects, protozoa, and insect larvae too which
can exist inside these plants. They seem immune to the paralyzing
effects of the digestive fluids. In fact, studies of this seeming contra-
diction have revealed that these few organisms may be protected from
digestion by unique antienzyme elements in their own bodies.

Naturalists also have reported that a strange, primitive, and
monkeylike animal, the insect-eating tarsier, most likely lives in har-
mony with nepenthes. The plants lure and catch the insects; the tarsier
pops around periodically to pluck out the morsels it desires.

Perhaps one nepenthes variety has evolved a way to fight back. *Nepenthes bicalcarata* has developed what looks like claws or hooks just below the lid. In these, especially, observers have found the remains of small animals, presumably small tarsiers which came to snatch a meal but were themselves snatched as a meal by the plant instead.

Depending on variety as well as soil and sun, nepenthes may range in color from dark, rich green to reddish or almost variegated color patterns.

From talks with horticulturists in charge of arboretums as well as amateur cultivators of carnivores, and through our own experience, we realize that more varieties of nepenthes are becoming available today. They do respond well to greenhouse and terrarium growing. Naturally, their special needs must be respected to produce satisfactory results.

Nepenthes, except for the seldom available, drier habitat types,

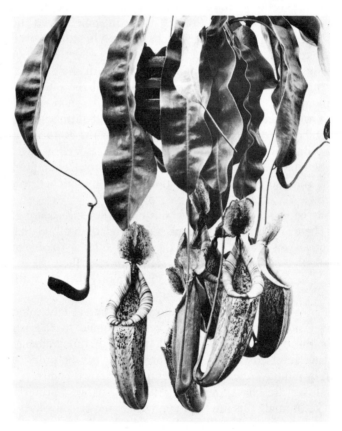

Nepenthes x dickinsoniana. (Courtesy Longwood Gardens)

require high humidity and ample moisture in the growing medium. A mixture of peat and sphagnum moss will provide the best growing mix. It holds moisture well, but also provides the adequate drainage which these plants do seem to need. We should point out that few plants can tolerate having their roots completely robbed of oxygen and air movement in the planting mix.

To assure adequate drainage, nepenthes should be placed in clay pots. The size depends on the variety, since some do mature to several feet tall. Place gravel in the bottom of the pot, then add your sphagnum moss. Since the plants do grow substantially, make certain to firm the moss to prevent the plant from toppling over as it gains stature.

If you can provide high humidity, from a misting system in a greenhouse, these plants can be grown in pots alone. Otherwise a large terrarium will provide that satisfactory moisture. When you consider that the moisture to permit the functions of the pitchers must be transported through the tendril which attaches the actual pitcher to the leaf, you can appreciate why high humidity is vital.

Nepenthes can be propagated by seed. They flower in captivity and seeds do form. Few beginners have had much success, since these plants are generally more difficult to grow than most other types of carnivores.

However, seeds are becoming more available from firms that specialize in carnivorous plants. Other hobbyists offer seeds in exchange for seeds or plants of other types of carnivores for their collections. *The Carnivorous Plant Newsletter* and Plant Oddities Club both offer this service to subscribers and members from time to time.

Plant nepenthes seed as you would the more easily grown pitcher plants. Use sphagnum moss mix as described in the care and culture chapter for your germinating medium. A soil heating cable or other bottom warmth under the propagating tray or pots is helpful to speed sprouting. We have found a 75° to 85° F. range works well. The moss should be moist at all times, especially as the plants begin to emerge, during those first few tender, critical weeks.

They also can be propagated from leaf cuttings. To do this, follow the same procedure recommended for New World pitcher plants.

Nepenthes deserve more attention. As more species become available, there will be greater opportunity to include these Far East oddities in school and home studies. Usually reliable sources for these plants are listed at the end of this book. Some firms have specimens available most times. Others offer different species periodically.

Foreign suppliers, from Indonesia to India, have special problems. Usually you must obtain import permits; they must get export permits. Then there is the matter of plant inspection before plants can be shipped. After all, plant inspection is necessary so that no insects which may be harmful to farm crops or shrubs and trees around your homes enter this country. That's why it does take time to obtain the more exotic species from lands far afield around the world.

If your curiosity is sufficient and and your patience and dedication at their peak, you can continue collecting such novel wonders as nepenthes year by year.

They may be nasty to their insect victims, but they do offer new horizons of exploration in this wonderful world of carnivorous plants.

Nepenthes rafflesiana growing in a hanging planter. (Courtesy Longwood Gardens)

Microscopic Carnivores

Up to now, we've concentrated on those carnivorous plants that can easily be grown, cared for, and watched as they demonstrate their marvelous skills.

But there are others that most people never see and few know exist. These too are quite carnivorous. Their strength is equal to the iron grip of the flytrap or the octopuslike clutch of the tentacled sundews.

Consider for a moment how powerful plants really are. The most slender root of a tree can work its way through the smallest crack in a rock ledge. It grows cell by cell. Eventually it can establish a sturdy roothold for that tree. You've seen this many times, trees jutting out from seemingly impenetrable rock cliffs along roadsides. Even grass, especially those weed grasses, prove their power many times. You may asphalt a drive, only to find some plants still forcing their way through what should be permanent impenetrable paving.

So it is not difficult to realize that some of the microscopic carnivorous plants also may have amazing strength. For example, there are many types of soil fungi in the plant kingdom. Admittedly they are quite primitive. Some of these have evolved a carnivorous pattern as the larger carnivorous plants have done. These fungi also have developed different ways to lure and catch their microscopic meals. Some have sticky discs to which tiny soil bacteria and nearly microscopic worms become stuck fast. Others have perfected nooses. They don't actually lasso their prey; instead, they beckon tiny soil eelworms, called nematodes, with their own versions of aromas and secretions. At least, that's how the agronomists and botanists we've talked with see it. As the nearly invisible worms crawl through the noose or hoop, these fungi sense they have a catch. The cells contract until they contact the unsuspecting worm. Thus, the fungi tightens the noose, effectively preventing the prey's escape. Watching this bizarre phenomenon requires keen eyes, a high-power microscope, and hours of patience.

Two of the smallest types of carnivorous plants belong to a group of fungi called molds. You can often find them in a bit of garden compost or the decaying leaf mold from a forest floor. When you place this sample under a microscope, there's lots to see. If lucky, you may have a portion of these carnivorous fungi. Their diet can also be revealed with proper magnification. Try 50 to 100 times at first. Once you locate the appropriate loops and snares you can zero in at higher magnification for a better picture of what is taking place.

Amoebas, nematodes, small crustaceans, and rotifers are all at work in typical compost or rotting leaf humus. They are the diet of these nearly microscopic carnivores. Nematodes seem to be the preferred food. Nematodes are tiny worms, about $\frac{1}{60}$ inch long. Some are helpful; others, like the golden nematode and potato wort nematode, are disaster when they invade farm fields. They stunt and kill the crops. You may have seen cabbage, broccoli, or cauliflower plants that are stunted, or dead. Chances are the root knot nematode has been at work. Fortunately, carnivorous fungi don't seem to discriminate. They eat all the nematodes they can catch.

One of the simple carnivorous fungi looks like a twig with tiny globules sticking out along its sides. Biologists have called it a "lethal lollipop." It is scientifically called *Dactylla asthenopaga*, quite a mouthful to say for such a tiny, primitive plant. When a nematode comes wiggling along and touches the knobs, it sticks to them. The more it squirms and wiggles the more its body touches the first and eventually other knobs. Scientists who have studied these primitive plants in action report that after the victim is effectively glued to the knobs, spearlike filaments are inserted into the worm's body. These serve to digest and absorb the "eel" dinner, leaving only the outer skin.

Another, even more ingenious fungus literally snares its meals. *Dactylaria gracilis* is a mold which grows nooses along its length. They may look like microscopic donuts, but these "donuts," of course, aren't to be eaten. They do the eating of any little nematodes or other creatures that crawl or move through the hole in the middle of the "donut." Each noose or snare has three cells which form the loop. They are joined to the stalk by a one-celled stem.

The outside of the snare seems lacking in response. Nematodes may touch it and provoke no reaction. But when the worm pokes its head through the noose, intending, I suppose, to crawl right through without a worry, then the fun begins. Fun for the plant, that is. Within a fraction of a second these three little cells change drastically. They swell several times in volume. In effect, this tightens the noose, putting a powerful hammer lock on the victim. As the victim struggles, prongs similar to those of the lethal lollipop are extended into the worm's

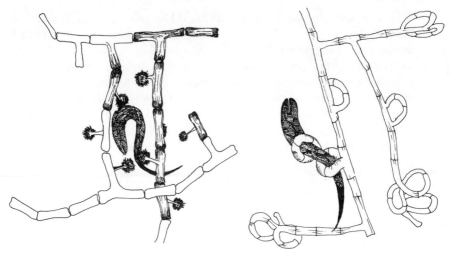

An eelworm caught by *Dactylla asthenopaga* (left) and *D. gracilis.*

body. They do the final job of absorbing and extracting the nutrients from the worm for the plant.

Although these carnivores prosper within soil containing nematodes, they can survive just on decaying organic matter. In this respect, they and other carnivorous fungi are like the larger insect-eaters. They can survive quite easily without a plentiful supply of insects or worms.

If your interests in carnivorous plants extend underground to those microscopic forms, you can actually stimulate these little critters to perform for you. The loop fungus is especially cooperative. Without nematodes in the soil, it merely grows steadily on without the loops. Or, at least, only a few tentative feelers. However, when you obtain some decaying organic matter or soil that contains nematodes, you can watch the plant go to work. By looking over samples gathered in your travels you can probably find some nematode soil, especially in gardens where cabbage, cauliflower, and their relatives show symptoms of root knot nematode damage. At this point, take the soil and mix it with a bit of water, enough to moisten the soil. Then, combine the soil with the organic matter which contains your carnivorous fungi friends. Day by day, through the microscope, you'll see the lethal loops begin to grow. From then on, keep an eye open. As the nematodes wiggle about, you'll eventually catch the action in the lens.

You can continue studies of these carnivorous fungi by cultivating them and microscopic amoebae and nematodes in agar solutions in petri dishes.

In a report printed in 1934, Charles Drechsler of the Bureau of Plant Industry outlined some interesting studies. You too can conduct similar tests. He noted that fungi on a natural substratum, their native habitat, often only reveal typical plant growth. But, he points out, very probably because nematodes and amoebae multiply actively and freely in agar plate cultures, the unique activities involving these animals as prey and the fungi can be more frequently seen. He identified two species of testaceous rhizopods, *Diffugia globulosa Duj.* and *Trinema enchelys Ehrenb.* These are both shelled protozoans, sluggish in movement. Not too sluggish to resent their entrapment by the fungi, of course. The plants appear to just wait, but as the protozoans move through the agar culture they become caught by the plants. On closer observation, the actual snaring can be seen, with the minute animals penetrated by the plant during the digestive process. All this must be viewed through the microscope, but in that normally unseen world the prowess of the carnivorous plants is apparent. Those that catch and eat animals are readily seen to increase in size. Probably in health, too, when they are better fed.

During his extensive studies, as reported in the *Journal of the Washington Academy of Sciences*, Vol. 24, no. 9, Mr. Drechsler noted that other minute plants also have the carnivorous habit. He mentions trichothecium, dactylaria, arthrobotrys, dectylella, and monacrosporium among the nematode-capturing species.

Other scientists in field and lab have expanded on these studies. It would seem that although most carnivorous plants exhibit their wonders aboveground, visible for all to see, nature's mysteries also abound in the soil, well hidden from the naked eye.

As you expand your explorations of the weird world of carnivorous plants, don't just look around. There may be just as many fascinating surprises down below, right beneath your feet. True, a microscope may be necessary. But on the trail of carnivorous plants, as we have been for many years, new worlds of wonder can be found in the most unlikely ways. And places too.

Carnivorous Curiosities
Among Smaller Families

There are a number of carnivorous plants that can't be classified with any of the large groups. One of these is *Cephalotus follicularis*. The rippled edge of its mouth with its hairy lining is rather nasty-looking. Not at all what you might expect from a plant that coaxes insects into its grasp. The trapping technique is similar to the purple pitcher plant, although they are not related. Another unusual attribute of this oddity is the fact that it has two types of leaves. One type seems to function only as a normal leaf. It is present for a large part of the year in its native habitat and may occur even after the few months that the hollow, insect-eating, and savage-looking leaves appear. Once these trapping pitchers begin to grow the plant forms dense clusters, spreading out into carpets of new plants.

Insects are lured into the poised mouths, then skid down the inward pointing hairs until they reach the waxier inner regions. Digestion takes place both by enzymes secreted by the digestive glands and bacterial action in the broth contained within the hollow pitchers.

Cephalotus is seldom available except through swaps of plants with collectors and fanciers. There's a reason for this: cephalotus is much more difficult to grow for any length of time. It prefers peat and sphagnum mixtures and ample moisture at all times. Most carnivorous enthusiasts find it difficult to keep for more than a year or two. Rarely do we ever hear of success in getting this stubborn plant to flower and set seeds. For those with time and patience, it deserves a place in the serious hobbyist's or specialist's collection. For beginners it is usually disappointing.

Another somewhat rare and touchy carnivorous curiosity is the heliamphora. It is known only in a native habitat on the remote Mount

Cephalotus follicularis is identified by the fluted edges on its pitchers and its tall flower stalks. (Courtesy Longwood Gardens)

Drivoa in Venezuela. We've tried to reach the area several times on trips to Caracas, but just couldn't contact the needed guide.

Heliamphora is distinguished by its wide funnel shape and has virtually no hood or flap. It too grows in mossy and boggy glades. Its funnels may rise almost six inches or longer to await their meals atop a mossy layer.

Although you might expect this plant would be flooded periodically in the rain forest atmosphere, it compensates for this excess moisture. It must, otherwise its meals would be washed away constantly. Instead of the hood or cap or flap common to most pitcher plants and even the nephenthes, the heliamphora has a vent or slit in the funnel. It works like a bathtub drain. When too much rain enters it flows out the upper drain, leaving the insect meals to be completely digested in the lower reaches of the funnel.

Drosera binata   *Drosera peltata*

Like other pitcher plants however, it does utilize a hairy upper zone inside the funnels graduating to a narrower waxy surface deeper inside. Blooms are attractive and delicate, rather showy white flowers on tall reddish stems. Although it seldom has performed its customary blooming in captivity, this plant is becoming more available from enthusiasts who have imported small numbers. According to qualified hobby growers and scientists at arboretums, heliamphora does respond to sphagnum and peat moss growing medium with constant high humidity.

Three types of sundews are included in this chapter rather than with the seemingly related sundews in their own chapter. These are natives of Australia and nearby areas; they are distinct from the others in growth pattern, and also a bit more difficult to obtain, so only specialists may be interested in pursuing their cultivation.

Drosera binata, the horseshoe sundew, is native to Australia, New South Wales, and New Zealand. Restrictive export requirements make it even more difficult to obtain the plants. Seeds are coming to the United States and growers should soon have limited quantities of seeds and plants for those who wish to observe these horseshoe shaped sundews in action. Fortunately, this sundew is prolific. An Australian friend looked at me curiously when I asked about this plant; he and others in Australia look upon the plant more as a weed than as anything of value.

This plant is usually short with a stem bearing the typical and characteristic horseshoe stems. These are lined with red-tipped tentacles which may remain greenish or pinkish if not given sufficient sun. As the plant grows, the horseshoe stem with its insect-catching tentacles may split several times until the top portion of the plant seems to be a maze of upward-pointing miniature horseshoes. It flowers annually, bearing white blooms on slender stalks. Culture of this sundew is the same as for native American or other types. Moist sphagnum moss over a layer of sand and peat is good. If it flowers, you can sow seeds. Otherwise, try propagating by root and leaf cuttings. It does have a better root structure than many of the small American sundews, so once you get it started right it should grow on for years.

Drosera peltata, the so-called bracketed sundew, is another strange sundew native to swamps and bogs of Australia. It has what appear as bursts of leaves covered with the typical red-tipped tentacles. These roundish leaves grow at various points along an upright stalk. To get a better understanding of this oddity, consider how typical round sundew leaves would look if you attached them every inch or so along a tall filiformis sundew's upright leaf.

The *D. peltata* has a vinelike growth pattern, reaching 6 to 10 inches tall. It bears white flowers in season. The plant is dormant during summer in its native Australia but begins new growth each autumn from its underground tuber. With luck, and you need it to grow this particular carnivore, the plant will send up its new shoots and flower by spring. After setting seed, in early summer, the vines droop and usually set roots as those portions of the stem touch the marshy ground again. That's much like the way black raspberries drop their branches to the ground and set new roots to produce new plants. Another example might be the spider plant that sets new plants at the ends of stems which can then be cut apart and repotted.

Also from that land of oddities that gave the world the duck-billed platypus and other animals found nowhere else on Earth comes *Drosera schizandra*, a sundew that seems to grow only in North Queensland and then in a restricted range near Mount Bartle Frese. It has broad leaves with a visible notch in the center. The plants have smooth, round leaves with the typical tentacles emerging also over them. They also grow in a radiating pattern much like the American sundews.

Before we leave Australia, there is one other carnivorous plant that captures the imagination. Not a sundew, although it may look that

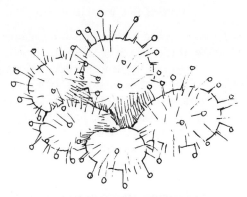

Drosera schizandra

Byblis gigantea exudes sticky digestive fluids whenever an insect lands on its long leaves. (Courtesy Longwood Gardens)

way, *Byblis gigantea* tops all the usual carnivorous plants in size. Growing from a woody base, it stands 2 feet tall or more and at first glance looks much like a tall filiformis sundew that had suddenly sprouted far beyond its normal size. The stalks are covered with drops of sticky fluid on what amounts to tiny tentacles which secrete the juices. This plant too can sparkle in the sun. In its native habitat *Byblis gigantea* looks much like a large, leafless shrub. When plentiful it even expands in a hedgelike growth pattern. What a plant to grow to keep stray cats and dogs from trespassing on your property.

Although it dines on insects of various size, this plant has a larger appetite, as you can well imagine. Its sheer size demands more nourishment. Small frogs, lizards, even young birds have been found stuck fast to the arms and branches of this hungry carnivore. In the few arboretums in which it has been cultivated the plant has not yet reached the size it achieves in its native land. However, the problem may be that we have become accustomed to maintaining higher moisture than it requires. In the wild, *Byblis gigantea* prefers fairly well-drained sandy soil, according to Australian botanists. They have propagated it from root cuttings and report that new plants are quite readily produced this way. Byblis also seems to store moisture in its roots to tide it over the drier times of the year.

In far-off Morocco and the dry soil of the hills of Portugal another strange carnivorous plant awaits its meals. Drosophyllum is a genus similar to the sundews, the drosera, but has none of its moist-soil-loving nature. Drosophyllum is a low, shrubby plant which produces long leaves bearing two different types of glands. One type is red and stalked, looking much like miniature toadstools around the stem. The other type is green and occurs right at the stem. Farmers have long picked this and hung it in the house to catch flies, a logical use. When an insect alights the stalked glands exude a sticky mucilage which functions as the catching mechanism. Although it appears that the tentacles move, they really don't. Insects are literally drowned by the secretions of the tall tentacles. Then the digestive glands, the green-colored ones along the stalk, secrete their digestive fluids to absorb the desired nutrients. A mosquito can be completely devoured in about 24 hours by this remarkably effective digestive process. Only the skeletal remains will be left to blow or wash away.

Drosophyllum is attractively colored by the reddish tentacles. Flowers are a bright yellow. At maturity the drosophyllum, or, as it is called in Portugal, slobbering pine, can grow well as a house plant, up to 18 to 24 inches tall. For culture, sources note that well-drained soil plus humid air around the plant is necessary for success. Although it grows well in its native areas, plants are seldom available in the United States and seed swaps are your best bet to obtain this carnivorous plant.

There are other carnivorous plants around the world, of course. Some undoubtedly have yet to be discovered lurking in the jungles of the Amazon or darkest Africa. Strange mountain passes of South America, the Orient, islands virtually unexplored in the South Pacific will probably yield new discoveries in the years ahead.

We hope that you, once your appetite for pursuing these botanical wonders is sufficiently whetted, may be among the plant explorers to discover new exotics. As you do, your name too may be added to the Latin identification for the plant, to record in history your discovery.

Search on. The reading list included in this book is quite complete. Libraries around the world no doubt hold more facts and figures. Once you're caught by carnivorous plants, you'll find many exciting new growing horizons across the world and friends to share your uncommon interest in these most marvelous of nature's plant wonders.

Experiments with Carnivorous Plants

You don't have to be a Charles Darwin or even a science student to enjoy growing and experimenting with carnivorous plants. They'll perform all sorts of fascinating feats as you watch them luring and catching insects.

In fact, every year across America and in many other countries around the world, tens of thousands of people from teachers and students to individual plant hobbyists, conduct an amazing range of tests and experiments with plants. Carnivorous plants receive far greater attention than most other plants. That's easy to understand—they are certainly among the world's most unique and fascinating plant wonders.

Perhaps you have never considered serious study of carnivorous plants and just prefer to enjoy watching them catch their meals. Well, why not try some experiments? You'll most likely surprise yourself with the discoveries you can make. After all, every new lesson learned helps expand your growing experiences and increases your own talents with green growing things. Be inquisitive! There are new horizons to explore.

Carnivorous plants are a mystery to millions of people. Many scientists are still probing their secrets, trying to discover why these plants developed their amazing capacity to lure, catch, and digest insects. And, how they actually do it. At some laboratories, researchers are actually attempting to synthesize the insect attractant of flytraps and other carnivorous plants. They theorize that if they are able to learn the exact chemical properties involved in the plant's insect-luring secretions they may be able to duplicate it chemically. That could lead to a new breakthrough in biological control of harmful insects on crops. Several researchers we interviewed pointed out that by using small amounts of that attractant at the edge of farm fields it may be possible to lure insects into a closed chamber in which they could be killed mechanically, chemically, or perhaps electrically. That development might eliminate the need to spray insecticides on certain food crops. This may sound like a long shot. Perhaps it is. But, in the field of horticultural and agricultural sciences, stranger things have been attempted and accomplished.

The secrets of carnivorous plants offer exciting challenges; these amazing plants hold real potential for new discoveries. Perhaps *you* may be the one to unlock another mystery in the wonderful world of plants. Whether you want to try just a few simple activities or really dig into advanced botanical studies, one thing is certain. These insect-eating oddities will have surprises in store for you.

For some twenty years I have been in touch with thousands of folks around the world who share my interest in these botanical wonders. We have swapped our knowledge by letter, phone, and in person. During my lectures and speeches in cities coast to coast, I've also interviewed thousands of people from all walks of life who are or have been involved in growing carnivorous plants of all types. Some of the most interesting experiences have been with students and their teachers as well as curious amateur plant enthusiasts. I can never tell where the next unusual project or discovery will come from.

Hundreds of people over the years have suggested many types of experiments, from simple feeding comparisons to far-ranging long-term studies. Some projects are suitable for simple show and tell activities for younger students. Others are more highly structured projects for high school and college level Science Fair studies and reports.

Space won't permit listing the hundreds of projects we have reviewed individually. Besides, half the fun is trying your own skill at experiments with plants and arriving at amazing discoveries by yourself. So we have put together in this chapter a broad range of suggested experiments. Some are just simple fun. Others will teach you a bit more of the little-known secrets of carnivorous plants, and the plant kingdom in general. Other ideas and recommended study programs will require more extensive knowledge or at least some dedicated work. How well you achieve the results you want will depend on many factors, from your previous experience to your ability to grow and apply experimental methods.

One thing is certain. These projects and these plants will assuredly put your powers of observation to the test. You'll most likely amaze yourself with what you can accomplish. We have grouped all the suggested fun and science study ideas in this chapter. As you develop your own project, either using one mentioned here or your own varia-

Three easily grown plants—the Venus flytrap in the foreground, miniature huntsman's horns in the left rear, and the purple or northern pitcher plant in the right rear—are a good starter group for study of different insect-catching techniques among carnivorous plants.

tions on it, you will find the details about the specific plants in the appropriate chapters of this book.

In addition, to help you with research and background, we have provided a fairly comprehensive reading list at the end of this book. Most libraries have useful information about many of the more common carnivorous plants in encyclopedias as well as botanical or horticultural books. Many libraries also can obtain the more detailed and sometimes technical references you may need for more advanced studies from their corresponding and associated libraries in your state. We also have provided lists of leading botanical gardens and horticultural societies which have helpful reference material available.

How much you dig into other sources depends of course on how extensive you wish to pursue your carnivorous plant investigations. Some friends have actually written their master and doctorate degree theses about insectivorous plants. That's a nice way to get a valued degree, writing about your hobby! Something you may wish to think about yourself, since a career in horticulture can be highly rewarding.

Career consideration aside, you may find not only fun, but awards and prizes available as you develop your research science talents and produce that first prize report for your local, regional, and even state Science Fair. Whatever your goals, we hope that some of the ideas that follow will be helpful as you expand your cultural horizons, in this fascinating field of carnivorous plants.

Venus Fly Traps

PROJECT 1

The easiest study project with these snappy little plants is a simple comparison of fed and unfed plants. We don't mean feeding hamburger, since, as you recall from the flytrap chapter, that will overtax the plants' digestive powers and can harm or kill them.

Materials—2 growing containers, 12 flytrap bulbs, planting material.

You'll need two identical, or quite similar, growing containers.

Plant several bulbs in these two separate containers. Use glass fish tanks, terrariums, or any suitable container that provides the proper high level of humidity these plants require. It is best to plant about four bulbs in each container so that you can observe differences more easily. True, you can notice the difference between just two plants, but several extras will provide a better chance of really graphic comparisons at the end of the test period.

Cover one container so that no insects can enter it. In effect, create a completely closed terrarium. As a closed chamber, the moisture will be taken in by the plants, given off in the normal process of growth form as condensation on the sides of the terrarium, and trickle back into the sphagnum moss planting material. This cycle in a properly planted, closed terrarium should be continuous and virtually self-sustaining.

Leave the second container open so that insects can be attracted in by the plants. You can also actually "feed" the plants in this container. Catch flies, small moths, mosquitoes, other tiny insects.

Use tweezers to carefully place the insect into a mature, open trap that is cocked and ready for its meal. Remember to touch the two trigger hairs so that the trap snaps.

Live insects are best, since their struggles cause the trap to close fully and stimulate the secretion of the digestive enzymes.

Don't overfeed the traps in the open container. Just one or two insects per plant are sufficient. You will need to add water periodically in the open terrarium or planter, since moisture does evaporate and escape.

Next, record when the plants snapped shut or were fed.

Then, watch them each day and record the date each snapped trap reopens.

When the plant has its jaws flexed again, let it catch its meals or try another feeding. At the end of a given time, one month or several, get out your ruler.

Measure the size of traps that have been eating, the number of traps on each of these plants, the condition of the plants. Compare those nourished plants with those that lived on photosynthesis alone.

Then, jot down your findings and write your report.

PROJECT 2

Materials—12 flytrap bulbs, planting container and moss, stop watch, movie camera.

Plant a dozen flytrap bulbs and allow them the required 8 to 10 weeks to mature and produce a nice display of hungry traps.

Get a stop watch and, if you wish, a home movie camera. Set up the camera and prepare the stop watch.

Next, take a pencil, piece of twig, tweezers, or other suitable tickler. Gently brush the trigger hairs of a cocked trap.

You'll have to be quick with the stop watch, since a mature, well-cocked trap with sufficient moisture in the cells can be mighty quick.

Next, try another open trap. With practice, you can get the feel of coordinating the watch and your triggering device.

Then, for comparison, let some of the plants dry out a bit. Cut down on the water provided. Then repeat your experiments.

If you wish, measure the amount of water provided different plants. The final testing can be done in just a few minutes once the plants are mature and ready to snap.

Again, write down the figures and you have a completed experiment that may just surprise you. How fast did these traps really snap shut? Amazing, but seeing and timing the action is proof positive.

When you use a movie camera, remember that you can achieve slow-motion effects by increasing the speed at which you take the pictures.

Then simply run the developed film at regular speed. One roll of 50 feet should be sufficient to record the results of the snapping action.

PROJECT 3

Materials—Microscope and slides, several flytrap bulbs, planting mix, containers.

For this project, you can grow several flytrap plants from bulbs to maturity. When they have caught several flies, cut off the closed traps. Remove one lobe or side of the trap and scrape some of the secretions around the insect off the trap and place on a viewing slide. Mount the cover over it and place under the microscope.

Then, look into a new world as you see the bacteria and accompanying ingredients of the digestive process taken from the inside of the trap.

You can take this study even further as some have already done. Keep records of when a trap snaps. Then, examine the contents through the microscope after the trap has been digesting the insect for 1 day, 2 days, 3 days, right up until the trap is ready to open naturally at the completion of its digestive process. What you find looking through the microscope can be a real eye-opener.

In fact, several years ago, one student in California grew 100 flytrap plants. He began a systematic study of the digestive system, opening traps and studying the contents and the process every few days. Result —a $1,000 scholarship for his winning Science Fair project. He found that the closed flytrap doesn't just rely on enzymes to digest the soft protein parts of the insect. What he learned was that the closed trap functions much like a cow's rumen or multiple-pouched stomach. Cows don't digest the grain and hay they eat. Microflora, millions of tiny active bacteria in the rumen, break down the hay and grain which in turn is utilized to nourish the cow. The great increase in bacteria that takes place within the closed trap also seems to function in this way, helping break down protein of the captured insect so that the enzymes can then further digest the materials and the bacteria themselves to nourish the plant.

Remember, this was not a scientist who devised the theory and conducted the experiments. He was a high school student.

PROJECT 4

Materials—Several flytrap bulbs, containers, various planting materials including sphagnum moss, peat moss, acid soil, sand.

This project is quite easy to do and proves several points. Simply plant the bulbs in the different planting mixtures. Provide the same growing conditions: moisture, opportunity to catch insects, etc. You can also grow them in completely controlled chambers to be more precise and avoid any distortions of results that might be cause by some traps catching insects while others don't.

At the end of a 6- to 10-week period, notice which bulbs have produced sturdy, healthy, large plants. You can even try combinations of vermiculite and peat, sphagnum moss and sand, almost any planting mix variation you want to evaluate.

For twenty years we have been periodically evaluating various growing mediums. One has consistently proved superior. This project will tell you rather graphically which is best.

PROJECT 5

Materials—Bulbs, planting mix, containers.

We pointed out in the chapter about flytraps that force feeding of hamburger and meat can be harmful. Granted, if you only give a trap the tiniest sliver it may be properly digested. Most people drop in what amounts to relatively large chunks of meat for the size of the trap, which can give the traps a bit of indigestion, and even kill them.

So, grow some plants. Let some lure and catch insects. Feed others flies or gnats you catch. Feed others hamburger or bits of liver. Remember to mark the pots or containers so you have a clear record of which plant was fed which meal.

Write down the dates, the day-by-day developments, and summarize your conclusions. Try feeding a trap ants. Then describe what effect the formic acid of the ant may have on the plant.

PROJECT 6

Materials—Large bulbs, planting containers, and moss.

Here's a propagating plan. As you develop your green thumb skills, you can probably produce new plants by bulb division.

Take several large bulbs and snap them in half or in thirds. Plant the divided parts and cultivate them carefully with sufficient humidity and 70° to 75° F. temperatures. Divide other bulbs by peeling off the various layers. Plant them also.

With luck and good care, you should be able to produce new plants by this simple method of bulb division.

PROJECT 7

Materials—Bulbs, containers, moss, fertilizer.

Another way to evaluate whether these unique plants need or can use fertilizer is to try giving them some. So, plant several containers.

When plants are almost mature, give several a teaspoon of liquid fertilizer based on the amounts recommended for that particular fertilizer for house plants. Give some plants a bit more or bit less. As a check, don't fertilize your control plants.

You'll find out quickly how these plants respond to chemical or even organic fertilizer. You can also achieve similar comparisons by trying tests using insect control sprays.

PROJECT 8

Materials—Bulbs, containers, planting material, knife.

This can be a variation on the bulb division, since it is a propagating technique. Remember, whenever you attempt any type of vegetative or asexual propagation, moisture is vital. Covered planters, terrariums, or just pots with plastic bags over them will do, but you must maintain the higher humidity to assist roots to form and the propagated cutting to grow.

Cut off leaves from the mature plants; some with large traps, some with smaller ones. Take half and dig the cut end in Rootone, the root-stimulating hormone. It is available at garden supply stores. Leave the other half of your cuttings alone.

Plant them all in the sphagnum moss. It helps to grind or shred the moss a bit to keep moisture around the cut end which you simply insert into the moss. Cover and await results.

This experiment is a bit touchy, since flytraps are difficult, in fact sometimes almost impossible, to propagate the first few tries. But with experience and perseverance, you'll get the feel of it and may be successful.

You can devise many other projects, from growing seeds after your plants have flowered to irradiating bulbs and seeds with various radioactive materials. Be imaginative. Be inquisitive.

You'll never know what you can really do and discover until you put your curiosity and talents to work and find out.

Sundews

Although not so active as the quick-snapping habits of the flytrap, sundews offer opportunities for a variety of study activities.

PROJECT 1

Materials—One or two varieties of sundews, containers, planting material.

Plant the sundews individually in the containers. Number them for easy reference. Provide the same humidity, light, care.

Then, let them catch whichever insects they can. If you have access to fruit flies, try planting sundews in covered containers and introducing several fruit flies.

If fruit flies are not available, cut a piece of banana and put bits on the planting material near each plant to help attract fruit flies.

Keep watch over your plants. Record when and how many insects are caught and what benefits the plants seem to obtain as they digest the insects.

PROJECT 2

Materials—Several plants, containers, mix.

Plant several sundews in different containers. Give some as much water as they can take. Keep the planting mix almost soggy. Give others adequate water so the mixture is moist to the touch, not wet.

The remaining plants should be allowed to become drier, without, of course, withholding so much water that you kill them.

Then, after a week or so, carefully examine the plants. With a hand lens closely study the amounts of "dew" secreted by the glands in the tentacles of the various plants. Record your results.

You can take this study a few steps further. Continue to grow the plants, maintaining the same culture conditions for each group.

Record the insect-catching ability of those kept soggy, those just moist, those growing in drier conditions.

PROJECT 3

Materials—Plants, various soil mixtures, containers.

This project is basically the same as the soil mix comparisons for the flytrap plants. You will, however, find that sundews do have a greater tolerance for other planting materials. That's natural, since their native habit range is worldwide. Find out what planting mixtures are best.

PROJECT 4

Materials—Several varieties of sundews, containers, planting mix, hand magnifying glass.

Grow a variety of sundews: rotundifolia, intermedia, filiformis. Then, count the tentacles on the arms. Next, either introduce fruit flies or encourage the native ones with bananas in the containers.

Day by day examine the plants. Count the insects trapped. Each day, re-examine the insects caught earlier to compare the progress of the various plants' digestive ability and speed.

There are many variations on this basic project, of course. Remember, however, that all plants should be cared for well with moisture in the planting mix and humidity around them adequate at all times, if you are to achieve the results you want. After all, you don't respond as well when you're not feeling up to par. Plants won't either.

PROJECT 5

Materials—Various sundews, containers, planting material.

Grow a variety of sundews just for fun. Then, on nice warm, sunny days, place them outdoors in your garden or a convenient location around the home. Let them catch insects to their hearts' content.

Obtain an insect identification book or field guide to common insects from the library. Watch your plants and try to identify the variety of insects they actually catch week after week.

We had a tray of tall sundews on the porch one year. One evening they seemed all aglow. In fact, so were the flytrap plants. On closer observation we discovered that they were dining on the first fireflies of the season that had begun emerging that week. A bit eerie, especially the flytraps that had a dim glow from inside the traps. That natural phosphorescence lasted until the traps reopened again and the husks of the fireflies were washed away by rain.

Pitcher Plants

These more passive plants won't provide the fact action of the flytrap or even the sundews. But, you have a greater range of them for comparative study and there are informative projects to be undertaken with them.

PROJECT 1

Materials—One or several varieties of pitcher plants, containers, planting mixture.

Grow the plants to maturity. That's the point when you'll have several fully grown pitchers, some half-grown, a few just emerging from the rhizomes. Let them catch insects outdoors or in a terrarium in which you introduce insects periodically.

When plants have thrived for a month or so, remove the largest pitchers, cutting them off right at the base. Next, take a sharp knife and/or single-edge razor blade and carefully slit the pitcher open. Use pushpins or straight pins to hold the pitcher open on a board.

What a sight! Try to identify the various insects. You'll probably be amazed at what the pitcher has eaten in the few short weeks it has been growing for you.

From this point, you can proceed with studies under the microscope to better identify the insect bits and pieces and even the bacterial culture which exists to assist the plant in its digestive process.

PROJECT 2

Materials—Pitcher plants, containers, mix, sharp knife, hand lens or magnifying glass.

Grow the plants for several months. Then cut off several mature pitchers. Slit the pitchers open.

Study the various areas carefully. From the spine-lined mouth of the purpurea to the striking coloration of the sweet trumpet's mouth, follow the insect's path along the various levels inside the pitcher.

Note the glands which secrete the aroma to attract the insects. Observe the glands which secrete the enzymes which help digest the insect.

You can even carefully slice away portions of these areas for further study under a microscope. Each part of the pitcher plant has a function to perform.

As you find and study these areas and observe the cross sections under lens or microscope, you can easily duplicate them on paper. Sketch what you see. Add your comments.

PROJECT 3

Materials—Several pitcher plant varieties, containers, mix.

Grow a variety of pitcher plants. As they grow, count the days from sprouting to maturity of the various pitchers of the different varieties. Then, sketch the plants, comparing in your descriptive notes the variations in their insect luring and catching methods. Best varieties include the northern or purple pitcher plant, miniature huntsman's horn, hooded pitcher plant, sweet trumpet, tall huntsman's horn.

Once you have a variety of plants growing, you can continue your studies right through flowering time.

PROJECT 4

Materials—Several varieties, containers, planting mix, small camel's-hair paintbrush, small scissors.

Once you have grown and studied the different varieties, try your hand at plant cross breeding or hybridizing. Many of these pitcher plants will cross with each other naturally. In fact, we have found over the years many natural hybrids in swamps and fields across the country.

You can duplicate nature at home or in the classroom. Once the plants flower, get out your scissors. Remove the pollen-bearing stamens from one variety while they are small so they don't begin to self-pollinate.

Then, use the camel's-hair brush to remove pollen from the other variety and transfer it to the pistils of the second variety from which you have removed the stamens.

You may not succeed the first try. But try again. It helps to have a dozen or more of each plant to increase your chances of success.

Keep records when you complete each phase of this work. Then, as seeds form and ripen watch the plants each week. When the seeds are dark and mature and ready to drop naturally, harvest your crop.

Next step is the most difficult, since it requires careful culture to germinate the seeds. Use a ground or milled sphagnum moss and sterile sand mixture, well moistened and covered by a plastic bag or other container to insure the needed high humidity.

With care and luck, you'll find the seeds begin to sprout in several weeks. Some authorities have suggested placing the seeds in the refrigerator for several weeks before planting. We have had success either way.

If your touch was right and the plants responded, you'll have achieved a most amazing result . . . a hybrid pitcher plant created by your own talents. That's an accomplishment.

PROJECT 5

Materials—Several plants, any variety, containers, mix, sharp knife.

Pitcher plants also can be propagated by rhizome division and with extra care by leaf cuttings. The northern or purple pitcher plant usually works well. So does the tall huntsman's horn, *S. flava.*

Simply wash all soil or moss from the roots and rhizomes. Then, with a sharp knife, cut the rhizome into pieces, being careful to leave roots on each portion.

Then, replant the cut rhizomes and keep moist while they take hold and begin to sprout new pitchers. We have found that the rhizome pieces should be ½ to 1 inch long. That means you can usually obtain 3 or 4 from a parent plant.

For leaf cuttings, simply remove the pitchers at the base and insert in your moist planting mix. You can try dipping some in the root-stimulating Rootone hormone powder according to the directions for using that product.

For practical purposes, we and others who have tried leaf propagating find about 25 to 50 per cent of the leaves, under adequate humidity and moisture, can sprout and eventually begin producing new plants. But, it does take care and luck. Rhizome propagating is easier.

Butterworts

Although these plants are rather passive, they do exhibit some action when insects are caught on the sticky leaf surfaces. Some varieties tend to curl their leaves when stimulated by insects more than others. That factor, however, may be more a response to the moisture conditions and general health of the individual plants than a variety differential.

PROJECT 1

Materials—Several butterworts, containers, mix.

Grow several butterworts and cover them. Leave the others open so that they can catch insects. This study can be combined with similar experiments using flytraps, sundews, pitcher plants, for simple observations and reports about the various plant habits.

PROJECT 2

Materials—Plants, containers, mix, hand lens.

Grow several butterworts and let them catch insects, or pop a few fruit flies and small gnats on the leaves. Remember, these are smaller plants and don't have the same appetites for larger insects as flytraps and pitcher plants.

Then, use your hand lens and carefully examine the insects as they are digested bit by bit. Record your observations. Keep a timetable of how long it takes the leaves to reduce the original insect into its final husk or skeleton stage. You might also consider using a ruler and measuring those leaves of well-fed plants for comparison with those which don't catch insects. Over a period of months you'll find some interesting differences in color, size, and general health of the plant.

PROJECT 3

Materials—Plants, containers, microscope.

Take the earlier projects a few steps further.

Remove a few leaves from the plants. Place them under a microscope. Observe and draw the areas you see, including the hairs, glands, and any insects caught. You might also try slicing cross sections of the leaf for even more detailed study.

PROJECT 4

Materials—Plants, containers, mix, fertilizer.

This can be combined with other carnivorous plant studies previously suggested. Just mix up your small amounts of fertilizer in water and compare the results when you apply it on the planting mix to those plants that are not subjected to chemical- or organic-based fertilizer solutions.

PROJECT 5

Materials—Several varieties, containers, mix.

Grow several of the different butterwort varieties. Keep a record of how long it takes for each to mature, how they catch insects, the size of mature plants, number of leaves, flowers, and other details. It is easy and fun, and from that point you can continue into more complex studies if you like.

Bladderworts

These mainly aquatic types of carnivorous plants have their own special techniques for catching their meals. In fact, they are sometimes so effective that they catch tiny, just-hatched fish. We have at present several gallon containers in which we're working on just such a fish-catching study.

Unfortunately it is not completed so we can't guarantee how the project will conclude. Besides, we're just a bit more partial and friendly toward fish, having raised tropical fish for years. Insect-catching is one thing. Sacrificing poor baby fish seems a bit more difficult to do, but then again, many projects are needed in the interests of scientific advancement.

Since few firms offer bladderworts, you may find it necessary to trudge along the edges of bogs and roadside ditches looking for these floating carnivorous plants. The illustrations in the bladderwort chapter provide you with a good guide to the general appearance of the plants. They bear tiny yellow, bluish, or white flowers on slender stalks, depending on variety.

In New Jersey you can often see thousands floating happily on the water of flooded cranberry bogs. They are also quite common along roadside drainage ditches in the South.

These plants are sometimes also available through scientific supply companies such as NASCO in Ft. Atkinson, Wisconsin, and Stansi-Fisher of Clear Lake, Wisconsin. Schools, of course, have access to them through their usual science supply sources.

PROJECT 1

Materials—Aquarium or gallon jars, plants, hand lens.

Grow several bladderworts for a few weeks in the same aquatic conditions in which you find them in their native habitat. Then, as they begin to thrive, remove several and place them on blotting paper or paper towels.

Use the hand lens and, if you wish, a single-edge razor blade. Explore their flotation devices and the tiny bladders along the stems.

With the hand lens, try to spot the minute aquatic insects and larvae which have been sucked into the bladders and are in the process of being digested.

PROJECT 2

Materials—Bladderworts, containers, pond water.

Grow the plants by floating them in the aquarium or jars. Introduce several cups of pond or stream water. When you obtain it, try to scoop up some of the nearly microscopic swimming insects. Most ponds have a lively population wriggling and swimming about.

Then, keep a careful watch over a few weeks. If you have keen eyes, you'll eventually be able to spot activity and compare the plants as they begin their eating and grow more healthy, week by week.

With a microscope, you might also try some peeks into the water and at the bladders of the plants. Be careful and calm as you slice into the bladders and open up these tiny compartments for study. It takes a steady hand for such minute plants, but you'll find yourself looking into an amazing world that you probably never knew existed.

PROJECT 3

Materials—Bladderworts, containers, tiny fish.

Begin by floating bladderworts in your aquarium. Then, introduce the tiny baby fish which you can scoop up with fine-mesh tropical fish nets from the edge of ponds in spring. We call them pinheads, since that's what they look like, just barely hatched and darting about. The most obvious characteristic is their head and eyes.

Hopefully, within about 10 weeks, you'll have some large bladderwort plants that have developed respectable-sized bladders. Keep a count of the fish. Note each day, since some may die and float to the top or drop to the bottom of the aquarium.

The reason to keep careful count is that you want to know when some pinheads are missing. From that point, begin the count of the bladders. Believe it or not, we and others have seen these tiny fish inside the bladders, sometimes with the tail tip still sticking out.

The sketch in the bladderwort chapter will show you what to look for in the typical ready-to-trap bladder, and one that had an aquatic mini-victim inside it.

Cobra Lilies

Strictly speaking, this name is really a popularized description given to the strangest of all native American pitcher plants. With its flared hood, reddish "tongue," and twisted pitchers, the plant does resemble a cobra snake if you use your imagination a bit.

We have cultured them for many years. During the past several years this plant has rapidly gained surprising popularity. Although it requires high humidity to prevent the tender pitchers from drying and is therefore harder to care for, it offers excellent study opportunities.

PROJECT 1

Materials—Plants, container, planting mix, knife.

Grow several cobra lily plants. Then, when some of the tallest pitchers are 15 or so inches high, remove them.

Carefully slice the entire pitcher open. Point out the translucent spots, the tongue, the mouth, the insect lure glands, and of course the various insects that have been caught by the plant. Add to your report with sketches and descriptions of what you find and identify.

PROJECT 2

Materials—Plants, several containers, planting mix.

Start with immature plants and grow them for several months or at least 6 weeks. Then, draw and describe all the parts, from juvenile pitchers to all the appendages in the fully mature plant. Compare its catching and eating habits, after watching it grow, with other types of carnivorous plants.

PROJECT 3

Materials—Plants, containers, planting mix.

Try feeding tests and fertilizing tests, too. Follow the same procedures as you would for the suggested project with various pitcher plants.

If you wish, and time permits, continue your studies with propagation efforts. Again, follow the methods for cutting off and planting individual pitchers. You'll find it is best to use the smaller to medium-size pitchers.

Those which are fully mature tend to have less remaining strength for setting roots, especially if they have already begun to dry out. That caution holds true for all pitcher plants.

When you make stem, leaf, tip cuttings for propagation purposes of any plant, the younger, tender, more succulent growth seems to have the greater strength and ability to catch hold and set new roots to become a plant itself.

You might also consider trying fed and nonfed comparisons using open and closed containers.

Or, simply for viewing pleasure, try just one beautiful cobra lily grown by itself in an attractive, tall, closed planter. Open it periodically so the plant can lure in its meals. We have seen cobra lilies thriving after several years, blooming year after year, in a 24-inch tall by 12×12-inch terrarium. It surely makes an amusing conversation piece on the coffee table when friends come to visit.

Combinations

Any of the studies suggested for individual plants or combinations can be accomplished in terrarium groupings too. In fact, most people prefer to grow and watch their lively little plant friends at work in a colorful display.

Here are some projects that can be successfully conducted when you elect to grow and study a variety of carnivores.

PROJECT 1

Materials—Flytraps, sundews, butterworts, several pitcher plants, container, and planting material.

Grow the plants together in a large fish tank or other sizable terrarium so they all have sufficient room to reach mature size without crowding. If the tank is low, select only those lower-growing types of pitchers. Huntsman's horns, sweet trumpets, and cobras tend to outgrow the usual 10- or 15-gallon tank. The result is curled tops as they meet the roof of their terrarium home.

Keep careful watch over your charges. Each week, note which have caught insects. Make a chart and list which seems to have the greatest insect-catching ability.

If you wish to compare open and closed containers, do so. You'll quickly find that those which can lure unwary insects to their snares will prosper.

PROJECT 2

Materials—Collect them all.

For a fascinating hobby that will astound your friends, begin an extensive collection. You can plant them in large tanks, a greenhouse, or individual terrariums and containers. Some friends have amassed beautiful collections. Several hobbyists in Japan, the United States, and Europe presently grow 50 to 150 different types of carnivorous plants.

These plants do have a way of growing on you. But then, some people raise hundreds of different cacti, others specialize in roses. One carnivorous collector in Japan has traveled the world, written and called us from around the globe. Although we haven't had the opportunity of seeing his collection firsthand, his photos prove his prowess in cultivating the largest collection of these plants we know of.

PROJECT 3

Materials—Various plants, containers, movie camera and/or slide camera.

Try your hand at photographing these marvels of nature. You already may be accomplished with a camera; if not, read up on photographic techniques. There are dozens of excellent books in your library or available at your local photo store.

Some of the newest movie cameras have stop action or slow motion capability. You simply set the camera for the number of frames desired per minute and let it run. You'll need to insure adequate light, of course. Every so many seconds the next few frames of the movie camera film are exposed. Bit by bit, hour by hour you can trace the growth and movements of your lively botanical specimens.

Or, if you prefer to work with slides, try various shutter speeds and exposure settings, as you trip the triggers of your powerful little flytraps. With practice, you can achieve some striking results.

Again, there are opportunities for applying your photographic skills to illustrate your study projects. Or, why not enter your astounding photos in a local photographic contest?

Try close-up shots, too. This requires special attachments or lenses, true, but with the fine cameras available today, you can focus in on your carnivorous plants to obtain truly amazing, prize-winning photographic results.

PROJECT 4

Materials—Various plants, containers, artificial lights.

You can achieve startling results with plants when you experiment with artificial light. Sunlight is preferred by these plants, naturally, but scientists have captured many of the most valuable rays of the plant growth light spectrum today in fluorescent and other types of bulbs. You can put these to work as part of a plant illuminating project.

Rather than get involved in complex terms from foot-candles to phototropism, we'll leave that in-depth research reading to you. Excellent books on growing plants under artificial light are available at schools and libraries everywhere.

What we suggest is that you apply some of these techniques to your carnivorous plant pets. Here are some ideas.

Suspend a twin-tube, 4-foot fluorescent reflector fixture over a plant stand by chains or on pulleys. Use a metal tray or other container with gravel in it on which you can place the potted plants. As an alternate, place the plants in open terrariums or pots which can be kept adequately moist daily. That's important, since even fluorescent lights give off some heat, and you don't want plants to dry and lose their vital bug-catching ability.

Make sure the fixture is properly grounded with a 3-prong plug. No sense shocking yourself when watering plants.

As you start plants from bulbs, roots, or rhizomes, lower the fixture to within 6 inches of the pots or trays. As plants grow you can raise the fixture.

Depending on how extensive you wish to make your studies, you can obtain a photographic light meter to accurately measure the light intensity in foot-candles. You can compare different light sources, from the Sylvania Gro-Lux lights to the Dura-Lite developments; their Vita-Lite and Natur-escent fluorescent tubes. Their newest Plant Lite is an individual bulb that fits into conventional light sockets. You can set up several of these for trials.

Consider testing carnivorous plant growth under different lengths of lighting periods, 8 hours, 12 hours, 14 hours per day.

You'll learn many things from these illuminating experiments. Plants will color up better, grow faster. For example, flytraps grown under Dura-Test lights were gaining bright red color inside the traps within

10 weeks. Some at 16 weeks were scarlet to dark maroon. Traps with less light remained pink or just green, although they did trap insects.

Newly planted bulbs grow to mature trapping plants in 8 weeks with 12 hours of artificial light per day. By 12 weeks most of 50 in one tray were flowering, too.

A twin tube unit 18 inches over one sundew bed of 500 plants helped turn them pinkish. Lowered to 6 inches over the plants, the Dura-Test Vita-Lites encouraged all the plants to reach a much more intense red. Held at 12 hours daily, they brought all plants into flower within 8 weeks of transplanting into the test beds.

There are dozens of other experiments with lights that can be devised. The lessons you learn will help you continue even more successfully with your carnivorous plant hobby for years to come.

Organizations, Periodicals, and a World List of Carnivorous Plants

Back in the early 1960s the interest in carnivorous plants began to expand. Perhaps this was a natural outgrowth of the increasing interest in all plants that developed as people began to want more knowledge of the wide range of plants they could grow in homes, offices, and schools.

It also can be attributed to the increasing awareness of these fascinating plants by naturalists and interested plant enthusiasts around the globe. Most likely the increased interest was helped along by the many different articles which have appeared in the recent decade focusing closer attention on the wonderful and exciting world of these insect-eating marvels.

Hundreds of articles have appeared in specialized as well as the popular press. From features in the *Reader's Digest* to *True* magazine, from newspaper columns and school publications, stories and articles have provided readers with heightened curiosity about the carnivores of the plant kingdom. That's only natural. These captivating plants do make news.

Two organizations can be credited with much of this long-needed effort to provide information on carnivorous plants. The Plant Oddities Club, through its worldwide membership, began in the 1960s from its former offices in Basking Ridge, New Jersey, to generate enthusiasm for exploring the mysteries of an ever-increasing range of carnivorous plants. Searching through old and current periodicals, the Club, now in Kennebunk, Maine, issues periodic bulletins to members accompanied by reprints from many popular and scientific journals.

Another group that has earned well-deserved attention for its part in building worldwide interest in carnivorous plants is the *Carnivorous Plant Newsletter*. It was launched in the early 1970s by co-editors D. E. Schnell of Statesville, North Carolina, and J. A. Mazrimas of Livermore, California. Together these two carnivorous plant enthusiasts have also helped to organize a global fraternity of carnivorous cultivators. Their periodic *Newsletter* is packed with names of subscribers from the far reaches of the planet.

Equally important, the brief but detailed articles provide new and accurate information about carnivores. Descriptions of growing and propagating methods, excellent photos and drawings by subscribers provide a steady supply of information.

In addition, these dedicated carnivorous collectors freely share their knowledge and their plants. There is a CPN swap shop for seeds, cuttings, plants. This enables other enthusiasts to obtain those rare species from growers in Japan, Europe, South America, and elsewhere in exchange for their own local types.

The *Newsletter* also provides details on subscribers' experiments, from the simple to the far-advanced electronic methods for testing plant responses.

But perhaps the greatest contribution of Mazrimas and Schnell has been their compilation of the ongoing world list of carnivorous plants. For several years they have spent untold dollars in postage corresponding with scientists and carnivorous plant hobbyists obtaining information about the locations and cultivation of carnivorous plants from the jungles of Brazil and Java, from the swamps of Canada and the bayous of Louisiana.

Since this list is expanding constantly it has become an important living thing itself. Updating is done periodically in the pages of the *Carnivorous Plant Newsletter*. Subscriptions are available for $2 per year within the continental United States, Canada, and Mexico; for $3 annually overseas. You can get details from either D. E. Schnell, Rt. 4, Box 275B in Statesville, North Carolina 28677 or Joseph A. Mazrimas, 329 Helen Way, Livermore, California 94550.

The Plant Oddities Club address is Box 94, Kennebunk, Maine 04043. The membership annual dues is $7, which includes the periodic Club bulletins, reprints of scientific and technical articles, and timely magazine and newspaper reports about carnivorous plants. In addition, the Club offers swaps of carnivorous plants as well as discounts through other growers for an increasingly wide range of plants from around the country and overseas as well.

With appreciation to Messrs. Schnell and Mazrimas for their untiring efforts we have included in this book, with their permission, a current list of the carnivorous plants found throughout the world. We have also included additions from my own and other sources gathered over the past few years, especially during preparation of this book.

As other carnivorous plant enthusiasts add to the needed store of knowledge, undoubtedly more detail about plant sources, ranges, and natural habitats will become available in the years ahead. We plan to add that information each time this book is revised for future publication.

World List of Carnivorous Plants

SARRACENIA

S. x ahlesii-alata x rubra—Alabama, Mississippi

S. alata—Alabama, Mississippi, Louisiana, Texas

S. alata x psittacina—Alabama, Mississippi

S. areolata-alata x leucophylla—Alabama, Mississippi

S. catesbaei-flava x purpurea—Virginia, North Carolina, South Carolina, Georgia, Florida, Alabama

S. chelsonii-purpurea x rubra—North Carolina, South Carolina, Georgia, Florida, Alabama

S. x courtii-psittacina x purpurea—Georgia, Florida, Alabama, Mississippi

S. x excellens-leucophylla x minor—Georgia, Florida

S. exornata-slata x purpurea—Alabama, Mississippi

S. flava—Virginia, North Carolina, South Carolina, Georgia, Florida, Alabama

S. flava x psittacina—Georgia, Florida, Alabama

S. formosa-minor x psittacina—Georgia, Florida

S. x gilpini-psittacina x rubra—Georgia, Florida, Alabama, Mississippi

S. x harperi-flava x minor—North Carolina, South Carolina, Georgia, Florida

S. leucophylla—Georgia, Florida, Alabama, Mississippi

S. minor—North Carolina, South Carolina, Georgia, Florida

S. mitchelliana-leucophylla x purpurea—Georgia, Florida, Alabama, Mississippi

S. mooreana-flava x leucophylla—Georgia, Florida, Alabama

S. oreophila—Georgia, Alabama

S. x popei-flava x rubra—North Carolina, South Carolina, Georgia, Florida, Alabama

S. psittacina—Georgia, Florida, Alabama, Mississippi, Louisiana

S. purpurea—Eastern North America

S. purpurea f. heterophylla—Newfoundland, Nova Scotia,

Massachusetts, New Jersey

S. *purpurea ssp. purpurea*—
Northern Range to New Jersey

S. *purpurea ssp. venosa*—Southern
Range from New Jersey

S. *x readii-leucophylla x
rubra*—Georgia, Florida,
Alabama, Mississippi

S. *x rehderi-minor x rubra*—North
Carolina, South Carolina,
Georgia, Florida

S. *rubra*—North Carolina, South
Carolina, Georgia, Florida,
Alabama, Mississippi

S. *rubra ssp. jones II*—RARE—
North Carolina, South
Carolina, Florida, Alabama,
Mississippi

S. *swaniana-minor x purpurea*—
North Carolina, South
Carolina, Georgia, Florida

S. *wrigleyana-leucophylla x
psittacina*—Georgia, Florida,
Alabama, Mississippi

PINGUICULA

P. *acuminata*—Mexico

P. *agnata*—Mexico

P. *albida*—Cuba

P. *alpina*—Europe, Asia

P. *antarctica*—Chile, Argentina

P. *balcanica*—Bulgaria, Yugoslavia,
Albania, Greece

P. *balcanica var.
tenuilaciniata*—Greece

P. *benedica*—Cuba

P. *caerulea*—North Carolina,
South Carolina, Georgia,
Florida

P. *calyptrata*—Colombia,
Ecuador

P. *caudata*—Mexico, Central
America

P. *casabitonana*—Cuba

P. *chilensis*—Chile, Argentina

P. *colimensis*—Mexico

P. *corsica*—Corsica

P. *crenatiloba*—Mexico,
Guatemala, Honduras, El
Salvador, Panama

P. *crystallina*—Cyprus

P. *cyclosecta*—Mexico

P. *elongata*—Venezuela, Colombia

P. *filifolia*—Cuba

P. *grandiflora*—Ireland, Spain,
France, Switzerland

P. *grandiflora f. pallida*—France

P. *grandiflora ssp. rosea*—France

P. *gypsicola*—Mexico

P. *heterophylla*—Mexico

P. *hirtiflora*—Italy, Eastern
Mediterranean

P. *hirtiflora f. pallida*—Italy,
Eastern Mediterranean

P. *hirtiflora var. louis II*—Italy,
Eastern Mediterranean

P. *hirtiflora var. megaspilaea*—
Italy, Eastern Mediterranean

P. *x hybrida-alpina x vulgaris*—
Russia, Finland, Austria

P. *imitatrix*—Mexico

P. *involuta*—Bolivia, Peru

P. *ionantha*—Florida

P. *jackii*—Cuba

P. *jackii var. parviflora*—Cuba

P. *kondoi*—Mexico

P. *leptoceras*—Switzerland,
Austria, Italy, France

P. *lignicola*—Cuba

P. *lilacina*—Mexico

P. *longifolia*—Spain, France, Italy

P. *longifolia ssp.
caussensis*—France

P. *longifolia ssp.
longifolia*—France

P. *longifolia ssp.
reichenbachiana*—France, Italy

P. *lusitanica*—Portugal, France,
Great Britain, Northern
Africa, Spain

P. *lutea*—Louisiana, Mississippi,
Alabama, North Carolina,
South Carolina, Georgia,
Florida

P. macroceras—Japan, U.S.S.R., Northwestern North America
P. macrophylla—Mexico
P. moranensis—Mexico, Guatemala, El Salvador
P. nevadensis—Spain
P. oblongiloba—Mexico
P. parvifolia—Mexico
P. planifolia—Florida, Mississippi
P. primuliflora—Alabama, Georgia, Florida, Mississippi
P. pumila—Texas, Louisiana, Alabama, North Carolina, Georgia, Florida, Bahama Islands
P. ramosa—Japan
P. x scullyi—grandiflora x vulgaris—Ireland, France
P. vallisneriifolia—Spain
P. vallisneriifolia luc, brevifolia—Spain
P. variegata—Siberia
P. villosa—Alaska, Canada, Sweden, Norway, Finland, U.S.S.R.
P. villosa f. albiflora—Northern Boreal region
P. villosa luc, ramosa—Northern Boreal region
P. vulgaris—Europe, Siberia, American Boreal region
P. vilgaris f. albida—Northern Boreal region
P. vulgaris f. bicolor—Northern Boreal region

NEPENTHES

N. alata—Philippines
N. albo-marginata—Malaysia, Sumatra, Borneo
N. ampullaria—Malaysia, New Guinea, Borneo, Sumatra
N. anamensis—Indochina
N. belli—Philippines
N. bicalcarata—Borneo
N. bongso—Sumatra

N. boschiana—Borneo
N. burbidgeae—Borneo
N. burkeii—Philippines
N. carunculata—Sumatra
N. cincta—Borneo
N. clipeata—Borneo
N. deaniana—Philippines
N. decurrens—Borneo
N. dunsiflora—Sumatra
N. distillatoria—Ceylon
N. dubia—Sumatra
N. edwardsiana—Borneo
N. ephippiata—Borneo
N. fusca—Borneo
N. geoffrayi—Indochina
N. globamphora—Philippines
N. gracilis—Borneo, Malaysia, Sumatra, Celebes
N. gracillima—Malaysia
N. gymanamphora—Sumatra, Borneo, Java
N. hirsuta—Borneo
N. x hookeriana-rafflesiana x ampullaria—Borneo, Sumatra, Malaysia
N. insignis—New Guinea
N. kampotiana—Indochina
N. khasiana—Assam, India
N. klossii—New Guinea
N. leptochila—Borneo
N. lowii—Borneo
N. macfarlanei—Malaysia
N. madagascariensis—Malagasy Republic
N. maxima—Celebes, Borneo, New Guinea
N. merrilliana—Philippines
N. mirabilis—Widespread distribution
N. mollis—Borneo
N. muluensis—Borneo
N. neglecta—Borneo
N. neo-guineesis—New Guinea
N. northiana—Borneo
N. oblanceolata—New Guinea
N. paniculata—New Guinea

N. *papuana*—New Guinea

N. *pectinata*—Sumatra

N. *pervillet*—Seychelles

N. *petiolata*—Philippines

N. *philippinensis*—Philippines

N. *pilosa*—Borneo

N. *rafflesiana*—Malaysia, Borneo, Sumatra

N. *rajah*—Borneo

N. *reinwardtiana*—Malaysia, Sumatra, Borneo

N. *sanguinea*—Malaya

N. *spathulata*—Sumatra

N. *spectabilis*—Sumatra

N. *stenophylla*—Borneo

N. *tentaculata*—Borneo, Celebes

N. *thorelii*—Cambodia

N. *tobaica*—Sumatra

N. *tomoriana*—Celebes

N. *treubiana*—New Guinea, Sumatra

N. *trichocarpa*—Sumatra, Borneo, Malaya

N. *truncata*—Philippines

N. *veitchii*—Borneo

N. *ventricosa*—Philippines

N. *vieillardi*—New Caledonia

N. *villosa*—Borneo

UTRICULARIA

U. *adpressa*—British Honduras, Venezuela, Guyana, Brazil, Trinidad

U. *albiflora*—Australia

U. *alpina*—South and Central America, West Indies

U. *amethystina*—Florida to Brazil

U. *andongensis*—Guinea to Zambia and Angola

U. *appendiculata*—Cameroon to Mozambique and Malagasy Republic

U. *arenaria*—Senegal to Ethiopia to Southwest and South Africa, Malagasy Republic, India

U. *aurea*—India to Australia, Japan

U. *aureomaculata*—Venezuela

U. *australis*—Tropical and South Africa, Europe, Temperate Asia to Japan, Australia, Tasmania, New Zealand

U. *baduleensis*—Tropical Africa, Malagasy Republic, India, Philippines, Australia

U. *benjaminiana*—Guyana, Surinam, Tropical Africa, Malagasy Republic, Trinidad

U. *bifida*—India to Australia, Japan

U. *biflora*—Eastern United States

U. *biloba*—Indochina, Thailand, Malaysia, Australia, New Guinea

U. *blanchetii*—Brazil

U. *bosminifera*—Thailand

U. *brachiata*—India

U. *bracteata*—Angola, Zambia, Zaïre

U. *bremii*—Europe

U. *breviscapa*—Cuba, Guyana, Brazil

U. *caerulea*—India to Australia, Japan

U. *calycifida*—Guyana, Venezuela, Surinam

U. *campbelliana*—Venezuela, Guyana

U. *canacorum*—New Caledonia

U. *capensis*—Malagasy Republic, South Africa

U. *capilliflora*—Australia

U. *cearana*—Brazil

U. *chiribiquetensis*—Colombia, Venezuela

U. *chrysantha*—Australia, New Guinea

U. *cornuta*—North America, Bahama Islands, Cuba

U. *cucullata*—South America

U. *cymbantha*—Zaïre, Mozambique, Zambia,

Botswana, Angola, South Africa, Malagasy Republic

U. delphinoides—Indochina, Thailand

U. dichotoma—Australia

U. dimorphantha—Japan

U. dunstani—Australia

U. endresii—Central America

U. erectiflora—British Honduras, Guyana, Venezuela, Brazil, Nicaragua, Colombia

U. evrardii—Indochina

U. exoleta—Africa to Australia, Japan, Spain, Portugal

U. fibrosa—Eastern United States

U. fimbriata—Colombia, Venezuela

U. firmula—Tropical and Subtropical Africa, Malagasy Republic

U. flaccida—Brazil, Paraguay, Colombia, Venezuela, Argentina

U. floridana—Florida, Georgia, North and South Carolina

U. foliosa—Florida to Argentina, Tropical Africa, Malagasy Republic, Galápagos Islands

U. fulva—Australia

U. geminiloba—Brazil

U. geminiscapa—Northeast United States

U. geoffrayi—Indochina, Thailand

U. gibba—United States

U. graminifolia—India to New Guinea

U. guyanensis—Trinidad, British Honduras, Guyana

U. hamiltoni—Australia

U. heterosepala—Philippines

U. heterochroma—Venezuela

U. hirta—India, Thailand, Indochina, Australia

U. hirtella—Central and South America

U. hispida—British Honduras to Northern Brazil, Trinidad

U. holtzei—Australia

U. hookeri—Australia

U. humboldtii—Venezuela Mount Roraima

U. Hydrocarpa—Cuba to Brazil

U. incisa—Cuba

U. inflata—Eastern North America

U. inflexa—Tropical Africa, Malagasy Republic, India

U. intermedia—Europe, Asia, North America

U. involvens—Malaysia

U. jamesoniana—Ecuador, Guyana, Venezuela, Colombia, Antilles, Peru, Costa Rica, Panama

U. juncea—Eastern United States, West Indies, British Honduras, Trinidad, Northern Brazil, Colombia, Venezuela, Guyana

U. kamienskii—Australia

U. kimberleyensis—Australia

U. kumaonensis—Himalayas

U. laciniata—Brazil

U. lasiocaulis—Australia

U. lateriflora—Australia

U. laxa—Argentina, Paraguay, Brazil

U. leptoplectra—Australia

U. leptorhyncha—Australia

U. limosa—Australia

U. livida—Ethiopia to Cape Province, Malagasy Republic, Mexico

U. lloydii—South and Central America

U. longeciliata—Surinam, Guyana, Venezuela, Colombia, Northern Brazil

U. longifolia—Brazil

U. macrorhiza—Temperate North America, Temperate East Asia

U. mannii—Bamenda Highlands, Cameroon Mountain, Cameroon; Gulf of Guinea

U. *menziesii*—Western Australia

U. *meyeri*—Brazil

U. *microcalyx*—Tropical Africa

U. *micropetala*—Western Tropical Africa

U. *minor*—Europe, Asia, North America

U. *minutissima*—India to Borneo, Australia, Japan, New Guinea

U. *monanthos*—Australia, Tasmania, New Zealand

U. *muelleri*—Northern Australia, New Guinea

U. *multicaulis*—China, Himalayas

U. *myriocista*—South America

U. *nana*—Brazil, Guyana, Surinam, Venezuela

U. *naviculata*—Brazil, Venezuela

U. *nelumbifolia*—Brazil

U. *neottioides*—Brazil, Colombia, Venezuela, Bolivia

U. *nephrophylla*—Brazil

U. *nigrescens*—Brazil

U. *novae-zelandiae*—New Zealand

U. *obtusa*—West Indies, South America, Tropical Africa

U. *ochroleuca*—Europe, Northwestern America

U. *odontosepala*—Malawi, Zambia, Zaïre

U. *odorata*—Indochina

U. *olivacea*—Eastern United States, Cuba, Venezuela, Guyana, Brazil, Surinam

U. *oliverana*—Venezuela, Amazonas (Northwestern Brazil), Colombia

U. *pentadactyla*—Ethiopia to Malawi and Rhodesia

U. *pierrei*—Indochina

U. *platensis*—Argentina, Uruguay

U. *podadena*—Southern Nyasaland, Northwestern Mozambique

U. *praelonga*—Brazil

U. *prehensilis*—Ethiopia to Rhodesia and Angola, South Africa, Malagasy Republic

U. *protrusa*—New Zealand

U. *pterocalycina*—Australia

U. *pubescens*—Guyana, Venezuela, Colombia, Brazil, Tropical Africa, India

U. *pulcherrima*—Trinidad, Brazil

U. *punctata*—India, Burma, Thailand, Borneo

U. *purpurea*—North America, Cuba, British Honduras

U. *purpureo-caerulea*—Brazil

U. *pusilla*—Central and South America, West Indies

U. *pygmaea*—Australia

U. *quelchii*—Venezuela, Guyana

U. *radiata*—North America

U. *reflexa*—Senegal to Southwest and South Africa, Malagasy Republic

U. *reniformis*—Brazil

U. *resupinata*—Eastern Canada and Eastern United States to British Honduras, Venezuela, Brazil

U. *reticulata*—India, Ceylon

U. *rigida*—Western Tropical Africa

U. *salwinensis*—China

U. *sandersoni*—South Africa

U. *sandwithii*—Guyana, Surinam, Venezuela

U. *scandens*—Tropical Africa, Malagasy Republic, Tropical Asia, Australia

U. *schultesii*—Colombia

U. *simplex*—Australia

U. *simulans*—Florida to Brazil, Tropical Africa

U. *singeriana*—Australia

U. *spiralis*—Tropical Africa

U. *stanfieldii*—Western Africa

U. *stellaris*—Tropical and Southern Africa, Malagasy Republic, Tropical Asia, Australia

U. *steyermarkii*—Venezuela

U. *striatula*—Tropical Africa,
India to New Guinea
U. *stricticaulis*—India
U. *subulata*—Nova Scotia to
Argentina, Tropical Africa,
Malagasy Republic, Thailand,
Borneo, Portugal
U. *tenuissima*—Trinidad,
Venezuela, Guyana,
Colombia, Northern Brazil
U. *tetraloba*—Western Africa
U. *trichophylla*—Brazil,
Venezuela, Guyana
U. *tricolor*—Brazil, Paraguay,
Colombia, Venezuela
U. *tridentate*—Southern Brazil,
Uruguay, Argentina
U. *triloba*—South America
U. *troupinii*—Tropical Africa
U. *tubulata*—Australia
U. *uliginosa*—India to Australia
U. *unifolia*—South and Central
America
U. *violacea*—Australia
U. *viscosa*—Trinidad, Venezuela,
Guyana, Brazil, British
Honduras
U. *vitellina*—Malaysia
U. *volubilis*—Australia
U. *vulgaris*—North Temperate
region including Europe,
Northern Africa, Temperate
Asia
U. *warmingii*—South America
U. *welwitschii*—Katanga, Rwanda,
and Burundi to Angola and
South Africa, Malagasy
Republic

DROSERA

D. *acaulis*—South Africa
D. *adelae*—Australia
D. *affinis*—Tropical Africa
D. *alba*—South Africa
D. *aliciae*—South Africa

D. *andersoniana*—Australia
D. *androsacea*—Australia
D. *anglica*—Europe, North
American Boreal region, Japan
D. *arcturi*—Australia, New
Zealand
D. *arenicola*—Venezuela
D. *ascendens*—Brazil
D. *auriculata*—Australia, New
Zealand
D. *banksii*—Australia
D. *bequaertii*—Central Africa
D. *binata*—Australia, New Zealand
D. *brevifolia*—North American
Boreal region
D. *bulbigena*—Australia
D. *bulbosa*—Australia
D. *burkeana*—South Africa
D. *burmanni*—Asia, Tropical
Australia
D. *caledonica*—New Caledonia
D. *calycina*—Western Australia
D. *capensis*—South Africa
D. *capillaris*—North and Central
America, Colombia, Brazil,
Guiana, Venezuela
D. *cayennensis*—Guiana, Brazil
D. *cendeensis*—Venezuela
D. *chiapensis*—Mexico
D. *chrysolepis*—Brazil
D. *cistiflora*—South Africa
D. *collinsiae*—South Africa
D. *colombiana*—Colombia
D. *communis*—Brazil, Colombia
D. *compacta*—Angola
D. *congolana*—Central Africa
D. *corsica*—Corsica
D. *cuneifolia*—South Africa
D. *dichrosepala*—Australia
D. *dielsiana*—South Africa
D. *drummondii*—Australia
D. *elongata*—Angola
D. *erythrorhiza*—Australia
D. *ferruginea*—Uruguay
D. *filicaulis*—Western Australia
D. *filiformis f. filiformis*—North
American Boreal region

D. *filiformis* f. *tracyi*—Gulf States,
United States
D. *finlaysoniana*—Vietnam
D. *flabellata*—Western Australia
D. *flexicaulis*—Tropical Africa
D. *gigantea*—Australia
D. *glabripes*—South Africa
D. *glandulingera*—Australia
D. *graminifolia*—Brazil
D. *hamiltoni*—Australia
D. *heterophylla*—Australia
D. *hilaris*—South Africa
D. *hirtella*—Brazil
D. *huegelii*—Australia
D. *humbertii*—Malagasy Republic
D. *incisa*—Cuba
D. x *hybrida-filiformis* x
intermedia—New Jersey,
United States
D. *indica*—Asia, Tropical
Australia, South Africa
D. *insolita*—Congo
D. *intermedia*—Europe, North
America, Guiana
D. *kaieteurensis*—Guiana
D. *katangensis*—Central Africa
D. *leucantha*—Southeastern
States, United States
D. *leucoblasta*—Australia
D. *linearis*—North American
Boreal region
D. *lovellae*—Australia
D. *macedonica*—Macedonia
D. *macloviana*—Falkland Islands
D. *macrantha*—Australia
D. *macrophylla*—Australia
D. *madagascariensis*—Malagasy
Republic, Tropical Africa
D. *maritima*—Brazil
D. *menziesii*—Australia
D. *metziana*—India
D. *microphylla*—Australia
D. *miniata*—Australia
D. *modesta*—Australia
D. *montana*—Brazil, Venezuela
D. *montana* var.
robusta—Venezuela

D. *montana* var.
roraimae—Venezuela
D. *myriantha*—Australia
D. x 'Nagamoto'-*anglica* x
spathulata—Japan
D. *natalensis*—South Africa
D. *neesii*—Australia
D. *neo-caledonica*—New
Caledonia
D. *nitidula*—Australia
D. x *obovata-rotundifolia* x
anglica—Asia
D. *occidentalis*—Western
Australia
D. *omissa*—Australia
D. *paleacea*—Australia
D. *pallida*—Australia
D. *parvifolia*—Brazil
D. *parvula*—Australia
D. *pauciflora*—South Africa
D. *peltata*—Australia, Japan,
Taiwan
D. *peltate* var. *foliosa*—Southern
Queensland, Australia
D. *peltata* var. *gracilis*—Southern
Queensland, Australia
D. *peltata* var. *lunata*—Japan,
Taiwan
D. *penicillaris*—Western Australia
D. *petiolaris*—Australia
D. *pilosa*—Cameroon, Kenya,
Tanzania
D. *planchonii*—Australia
D. *platypoda*—Australia
D. *platystagma*—Australia
D. *prolifera*—Queensland,
Australia
D. *pulchella*—Australia
D. *pusilla*—Venezuela
D. *pycnoblasta*—Australia
D. *pygmaea*—Australia, New
Zealand
D. *ramellosa*—Australia
D. *ramentacea*—South Africa
D. *regia*—South Africa
D. *rosulata*—Australia

D. *rotundifolia*—Northern
Hemisphere
D. *rotundifolia* x *intermedia*—
United States
D. *rubiginosa*—New Caledonia
D. *schizandra*—Australia
D. *scorpioides*—Australia
D. *sessilifolia*—Brazil, Guiana
D. *sewelliae*—Australia
D. *spathulata*—Australia, New
Zealand, Japan
D. *spiralis*—Brazil
D. *squamosa*—Australia
D. *stenopetala*—New Zealand
D. *stolonifera*—Australia

D. *stricticaulis*—Australia
D. *sulphurea*—Australia
D. *subhirtella*—Australia
D. *tenella*—Argentina
D. *thysanosepala*—Australia
D. *tomentosa*—Brazil
D. *trinervia*—South Africa
D. *umbellata*—China
D. *uniflora*—South America
D. *villosa*—Brazil
D. *whittakerii*—Australia
D. *whittakerii* var.
praefolia—South Australia
D. *zonaria*—Australia

Smaller Genera

HELIAMPHORA

H. *heterodoxa*—Mount
Ptari-Tepui, Venezuela
H. *macdonaldae*—Mount Duida,
Venezuela
H. *minor*—Mount Auyan-Tepui,
Venezuela
H. *nutans*—Mount Roraima,
Venezuela
H. *tatei*—Mount Duida,
Venezuela
H. *tyleri*—Mount Duida,
Venezuela

Aldrovanda vesiculosa—Europe,
India, Japan, Africa

Byblis gigantea—Australia
Byblis liniflora—Australia

Cephalotus follicularis—Australia

Chrysamphora californica—
California, Oregon

Dionaea muscipula—Eastern
United States

Drosophyllum lusitanicum—Spain

Carnivorous Plants
on Display

With the increased interest in carnivorous plants has come a welcome up-surge in displays of these botanical curiosities at arboretums, botanical gardens, colleges and universities. Some private collections also are available for viewing upon written request to the individuals concerned.

There are most likely many other displays, some large, others small, at various public and private gardens around the country. We have included here those gardens and displays which have had and usually have a current display of carnivorous plants. Before traveling long distances, it is best to write to the curator or chief horticulturist to determine whether the display is still on.

I am always anxious to obtain information about other display gardens, collections, or even specimens of the more exotic carnivorous plants. Any leads to help me obtain additional research publications, papers, and books, will be most appreciated.

Periodic or Permanent Plant Displays

Brooklyn Botanical Gardens, Prospect Park, Brooklyn, New York
Bronx Botanical Garden, also known as New York Botanical Garden, Bronx, New York
California State University, Fullerton, California
California State University, Humboldt, California
University of California at Berkeley, Berkeley, California
Columbia Zoological Park and Botanical Garden, Columbia, South Carolina
Longwood Botanical Gardens, Kennett Square, Pennsylvania
Cornell University Arboretum, Ithaca, New York
Los Angeles State and County Arboretum, Arcadia, California
Missouri Botanical Gardens, Tower Grove Avenue, St. Louis, Missouri
Phipps Conservatory, Schenley Park, Pittsburgh, Pennsylvania
San Francisco Conservatory, Golden Gate Park, San Francisco, California
University of North Carolina, Chapel Hill, North Carolina
State University of North Carolina, Raleigh, North Carolina
U. S. Department of Agriculture Research Station, Beltsville, Maryland
Denver Botanical Gardens, Denver, Colorado

IN CANADA

Montreal Botanic Garden, Sherbrook Street, Montreal, Quebec

IN IRELAND

National Botanic Gardens, Glasnevin, Eire

IN ENGLAND

Royal Botanic Garden, Edinburgh, Scotland
Royal Botanic Gardens, also known as Kew Gardens, Surrey
University Botanic Gardens at Cambridge and Oxford

PRIVATE COLLECTIONS

Randall Schwartz, Box 283, Lenox Hill Station, New York, New York
Plant Oddities Club, Kennebunk, Maine
James Pietropaolo, Canandaigua, New York

Leading Horticultural Libraries and Information Centers

CALIFORNIA

California Academy of Sciences
    Library
Golden Gate Park
San Francisco, 94118

California State Polytechnic
    College Library
3801 West Temple Avenue
Pomona, 91768

Forest History Society
Box 1581
Santa Cruz, 95060

Los Angeles State and County
    Arboretum Library
301 North Baldwin Avenue
Arcadia, 91006

Rancho Santa Ana Botanic
    Garden Library
1500 North College Avenue
Claremont, 91711

COLORADO

Denver Botanic Gardens, Helen K.
    Fowler Library
909 York Street
Denver, 80206

DISTRICT OF COLUMBIA

Dumbarton Oaks, Garden Library
1703 32nd Street, NW
Washington, 20007

U.S. National Arboretum Library
U.S. National Arboretum
Washington, 20002

FLORIDA

Fairchild Tropical Garden,
    Montgomery Library
10901 Old Cutler Road
Miami, 33156

Hume Library
University of Florida
Gainesville, 32601

GEORGIA

Callaway Gardens
Pine Mountain, 30822

ILLINOIS

Chicago Horticultural Society
Library
116 S. Michigan Avenue
Chicago, 60603

Lake Forest Library
360 Deerpath Avenue
Lake Forest, 60045

Morton Arboretum, Sterling
Morton Library
Lisle, 60532

INDIANA

Purdue University,
Forestry-Horticulture Library
Lafayette, 47907

MARYLAND

National Agricultural Library,
USDA
Intersection 1-495 and U.S. 1
Beltsville, 20705

MASSACHUSETTS

Arnold Arboretum
The Arborway
Jamaica Plain, 02130

Massachusetts Horticultural
Society
300 Massachusetts Avenue
Boston, 02115

Oakes Ames Orchid Library
22 Divinity Avenue, Room 109
Cambridge, 02138

Old Sturbridge Village Library
Sturbridge, 01566

University of Massachusetts,
Morrill Library
Amherst, 01002

Wellesley College Library,
Biological Sciences
Sage Hall, Wellesley College
Wellesley, 02181

Worcester County Horticultural
Society
30 Elm Street
Worcester, 01608

MICHIGAN

Michigan Horticultural Society
The White House, Belle Isle
Detroit, 48207

MINNESOTA

University of Minnesota, St. Paul
Campus Library
St. Paul, 55101

MISSISSIPPI

Mississippi Agricultural
Experiment Station
Stoneville, 38776

MISSOURI

Missouri Botanical Garden Library
2315 Tower Grove Avenue
St. Louis, 63110

National Council of State Garden
Club, Inc.
4401 Magnolia Avenue
St. Louis, 63110

NEW HAMPSHIRE

University of New Hampshire
Biological Sciences Library
Kendall Hall
Durham,   03824

NEW JERSEY

Rutgers University, College of
Agriculture and
Environmental Science
New Brunswick,   08903

NEW YORK

Cornell University, Albert R.
Mann Library
Ithaca,   14850

Garden Center of Rochester
5 Castle Park
Rochester,   14620

Highland Park Herbarium Library
Monroe County Parks
375 Westfall Road
Rochester,   14620

Horticultural Society of New York,
Inc.
128 West 58th Street
New York,   10019

New York Botanical Garden
Library
Bronx, New York   10458

New York State Agricultural
Experiment Station Library
Geneva,   14456

State University of New York,
Walter C. Hinkle Memorial
Library
Alfred,   14802

NORTH CAROLINA

North Carolina State University
D. H. Hill Library
Raleigh,   27607

University of North Carolina,
Botany Library
301 Coker Hall
Chapel Hill,   27514

OHIO

American Rose Society Lending
Library
4048 Roselea Place
Columbus,   43214

Garden Center of Greater
Cleveland Eleanor Squire
Library
11030 East Blvd.
Cleveland,   44106

Holden Arboretum Library
9500 Sperry Road, Kirtland, P.O.
Mentor,   44060

Kingwood Center Library
Box 1186
Mansfield,   44903

Ohio Agricultural Research and
Development Center Library
Wooster,   44691

Youngstown Garden Center
123 McKinley Avenue
Youngstown,   44509

OREGON

Oregon State University Library
Corvallis,   97331

PENNSYLVANIA

Hunt Botanical Library
Carnegie-Mellon University
Pittsburgh, 15213

Longwood Gardens Library
Kennett Square, 19348

Morris Arboretum of the
University of Pennsylvania
9414 Meadowbrook Avenue
Philadelphia 19118

Pennsylvania Horticultural Society
325 Walnut Street
Philadelphia, 19106

Pennsylvania State University,
Agricultural and Biological
Sciences Library
University Park, 16802

Temple University, Ambler
Campus Library
Meetinghouse Road
Ambler 19002

SOUTH DAKOTA

South Dakota State University,
Lincoln Memorial Library
Brookings, 57006

TEXAS

Texas Research Foundation,
Lundell Rare Book Library,
Suggs Library
Renner, 75079

VERMONT

University of Vermont, Guy W.
Bailey Library
Burlington, 05401

WASHINGTON

University of Washington
Arboretum
Seattle, 98105

WEST VIRGINIA

Wheeling Garden Center Library
Oglebay Park
Wheeling, 26003

ONTARIO, CANADA

Civic Garden Centre Library
777 Lawrence Avenue, E.
Don Mills, 404

Royal Botanical Gardens
Box 399, Sta. A.
Hamilton, 20

Sources
for Plants

During my years of studying, growing, and propagating carnivorous plants, I have contacted thousands of individuals around the globe. Many just grow a few plants for their own pleasure. Some develop extensive collections. Few have sufficient plants available to offer them for sale in quantity.

Several firms, however, have become deeply involved in the growing and propagating of these oddities. They offer catalogs and price lists so that you can order many of these botanical curiosities. Here's a current list of those firms:

Armstrong Associates, Inc., Kennebunk, Maine 04043, has an illustrated, informative catalog with many types, including Venus flytraps, sundews, butterworts, cobra lilies, and pitcher plants individually, as sets, and complete terrarium kits. Catalog 25¢

Insectivorous Botanical Gardens, 1918 Market Street, Wilmington, North Carolina 28403, offers flytrap bulbs and a variety of sundews, butterworts, and some pitcher plants. Catalog 25¢

King's Park and Botanical Garden, Perth, Western Australia, 6005, often has seeds of various Australian species of carnivorous plants.

Marcel Lecoufle, 5, rue de Paris, 94470 Boissy St. Léger, France, has a catalog offering several species of American plants as well as the Asian pitcher plants and some other exotics too.

Randall Schwartz, Box 283, Lenox Hill Station, New York, New York 10021, has been studying and breeding plants in recent years. He may have some plants or seeds available from time to time.

Brown Bulb Ranch, Capitola, California, has flytraps and a few other types available periodically, mostly seasonal.

Peter Paul's Nursery, Canandaigua, New York 14424, has offered carnivorous plants for about twelve years. They have a catalog listing the various types for 25¢.

Harold Welch, 266 Kipp Street, Hackensack, New Jersey 07601, has recently joined the suppliers of carnivores. His catalog is 25¢.

Edmund Scientific Company, 101 East Gloucester Pike, Barrington, New Jersey 08007, offers several plants individually and excellent kits. Their extensive science supply catalog is free.

Plant Oddities Club, P.O. Box 94, Kennebunk, Maine 04043, has been involved in detailed study of carnivorous plants for many years. This organization offers free literature about membership and special member discounts on carnivorous plants. They publish a unique, illustrated catalog each year, available for 25¢. The Plant Oddities Club also has a list of members willing to swap carnivorous plants with others. It is available to members.

Vaughan's Seed Company, 5300 Katrine Avenue, Downers Grove, Illinois 60515, lists Venus flytraps and other carnivorous plants available to garden centers and florists.

The Carnivorous Plant Newsletter, 329 Helen Way, Livermore, California 94550, has source lists among subscribers who swap and exchange plants.

Bibliography and Reading List

During the past several years, carnivorous plants have gained much well-deserved publicity and popularity. Many articles and information releases from the Plant Oddities Club have appeared in newspapers and magazines around the country. In addition, some stories about the club on the Associated Press wire service also gave broad coverage to the topic of these botanical wonders. Articles in a variety of magazines by botanists as well as professional writers have contributed to the renewed interest in these unusual plants.

We have included in this book a combination reading list and bibliography. I have written numerous articles myself and many other writers have contributed to the information about these plants. Some have added to the folklore as well. Among these sources are popular articles, technical bulletins, and complete dissertations. We have tried to compile as extensive a list as possible, including dates of publication of articles for easier reference through your local libraries where possible.

Undoubtedly there are valuable references and material which I have missed in my research for this particular book, although I have been collecting information for twenty years. There are some sources which may be excellent, but they are published in foreign languages. (My French is rusty and my Japanese nonexistent.) However, for those of you who wish to pursue all avenues, I have listed publications in other tongues. The photographs in some are worth seeing. Perhaps you also read the languages.

Bibliography

"Action Potentials Obtained From Venus Fly Trap" by C. Stuhlman and
  E. Darden in *Science*, 1950

"American Droseras in Sidney" by Stephan Clemensha, from *Carnivorous Plant Newsletter,* reprinted by permission by Plant Oddities Club, 1974

"Butterworts and Bladderworts" by R. E. Stauffer in *Journal of the New York Botanical Gardens,* 1950

Carnivorous Plants—Rare Plants That Eat Insects by Allan A. Swenson, Plant Oddities Club, 1973

Carnivorous Plants Provide Fascinating Science Projects by Allan A. Swenson, Plant Oddities Club, 1974

Carnivorous Plant article in *BioScience,* March 1965

"Carnivorous Plants Bulletin" by Carolina Biological Supply Company, 1965

Carnivorous Plants and the Man-Eating Tree by Sophia Prior, Field Museum of Natural History, Chicago, 1939

"Carnivorous Plants of the Illawarra Area" by Brian Whitehead, *Carnivorous Plant Newsletter,* reprinted by Plant Oddities Club, 1974

"Data on *Sarracenia flava*" by S. T. McDaniel as a Ph.D. dissertation, Florida State University, 1966

"Development of *Dionaea muscipula*" by Cornelia M. Smith in the *Botanical Gazette* of 1929

"Distribution of the Venus Fly Trap" by W. C. Coker, *Journal of the Elisha Mitchell Scientific Society,* July 1928

"Do Plants Have Feelings?" by Margaret Ronan, *Voice,* March 1974

"Drosera in the Southeastern United States," a report by L. R. Shinners, 1962

"Drosera in Eastern North America" by F. E. Wynne, Bulletin of the Torrey Botanical Club, 1944

"Evidence of the Hybrid Origin of *Drosera anglica*" by C. E. Wood, Jr., 1955

"Flowers That Kill to Eat" by Michael A. Godfrey in *National Wildlife,* August 1972, reprinted by permission by Plant Oddities Club, 1974

"Fly in the Sundew" by Terry Ashley and Joseph F. Gennaro, Jr., in *Natural History* magazine, December 1971

"Glittering Grabber" by Patricia Hollan in National Geographic School Bulletin, November 1973

"Identity of *Drosera brevifolia*" by Carroll E. Wood, Jr., *Journal of the Arnold Arboretum,* 1966

Illustrations of North American Pitcher Plants by Mary V. Walcott, with associated notes by Edgar T. Wherry and Frank M. Jones, Smithsonian Institution, Washington, D.C., 1935

Insect Paralyzing Agents from the Pitcher Plant Sarracenia flava, by Dr. D. Howard Miles, Mississippi State University, 1975

"Insect Trapping Plants" by Virgil N. Argo, *National History* magazine, 1964

"Life on the Sticky Sundew" by Thomas Eisner, *Natural History* magazine, 1974

"Man-Eating Trees," Willard N. Clute in *American Botanist*, April 1925

"Mechanics of Movement in *Drosera rotundifolia*" by J. D. Hooker in the Torrey Botanical Blue Bulletin, 1917

Modified Floral Parts of Dionaea by Ted A. Minton and Dr. Donald B. Jeffreys, a Plant Oddities Club reprint, 1974

"Natural Hybrids in the *Genus Sarracenia*" by Clyde R. Bell in the *Journal of the Elisha Mitchell Scientific Society*, 1952

"Nature's Switch, Plants That Eat Bugs" by R. Eliot Stauffer in *Conservation*, August 1972 reprinted with permission by Plant Oddities Club, 1974

Nature's Wonders by Charles L. Sherman, Nelson Doubleday, 1956

"Physical Analysis of Opening and Closing Movements of the Venus Fly Trap" by Otto Stuhlman in the Torrey Botanical Club Bulletin, 1948

"Pisciborous Plants" by G. E. Simms, in Bulletin of the U. S. Fish Commission, No. 4 in 1884, a real oldie source.

"Pinguicula in the Southeastern United States" by R. K. Godfrey and C. E. Wood, Jr., 1957

"Plants That Eat Animals," *Science World*, September 25, 1958

"Plants That Eat Insects" by Jean George, originally in *Au Grand Air*, 1962, reprinted in *Reader's Digest*, February 1963

Plants That Eat Insects by Allan Swenson, Terrarium Topics, 1974

"Plants That Eat Insects" by P. A. Zahl in *National Geographic*, 1961

"Potentials Developed in Venus Fly Trap Fundamental Action" by Otto Stuhlman in the *Journal of the Elisha Mitchell Scientific Society*, 1950

"Reappraisal of *Utricularia inflata* and *Utricularia radiata*" by R. K. Godfrey and G. W. Reinert in *American Journal of Botany*, 1962

"Sacrificed to a Man-Eating Plant" by B. H. William in *American Weekly*, September 1920

"Some Aspects of the Ecological Life History of *Sarracenia purpurea*" by A. J. Mandossian as a doctorate thesis at Michigan State University in 1965

"Synopsis of Pinguicula in the Southeastern U. S." by H. L. Stripling and R. K. Godfrey in *American Midland Naturalist*, 1961

"The Memory of the Venus Fly Trap" by Stephen E. Williams, originally in the *Carnivorous Plant Newsletter*, 1973, reprinted with permission by Plant Oddities Club, 1974

"The Only Known Fish-Eating Plants, Utricularia, the Bladderworts" by E. Guger in *Scientific Monthly*, 1947

"The Genera of Sarraceniaceae and Droseraceae in the Southeastern United States" by Carroll E. Wood, Jr., *Journal of the Arnold Arboretum*, 1960

Travels and Researches of a Naturalist in Borneo by Odoardo Beccari, London, 1904

"Venus Fly Trap, the Plant That Eats Insects" by Maurice Franz, *Organic Gardening and Farming*, 1972

"Venus Fly Trap" by Frank A. Montgomery, Jr., in *Raleigh News and Observer*, December 25, 1952

Other Books About Carnivorous Plants

Insect Eating Plants by Lynn and Grey Poole, T. Y. Crowell, 1962

Insectivorous Plants by Charles Darwin, John Murray, 1893, reprinted by AMS Press, Inc. 1971

Carnivorous Plants by Randall Schwartz, Praeger Publishers, 1974

A Fly Trap on Venus by C. B. Woodcock, Privately published, 1960

The Carnivorous Plants by Francis E. Lloyd, Chronica Botanica Company, 1942

The World of Carnivorous Plants by James and Patricia Ann Pietropaolo, R. J. Stoneridge, 1974

Index